THE 3Qs: QUANTUM QUANDARIES AND QUACKERY

The Philosophy of Science Applied to Modern Physics

1st Edition

Table of Contents .p. 4

By Douglas E. Reinhardt
Ph.D., UNC-Chapel Hill

Key Subjects

An Alternative Physics, A Skeptical Approach to Modern Physics, Philosophy of Physics, Philosophy of Science, Epistemology, Anti-relativity, Quantum Mechanics, Quantum Mysticism, String Theory, Scientific Realism, Metaphysics, Cosmology, Critical Thinking, Metacognition, Research Methods, Logic, Rational-Empiricism, Logical Positivism

Key Question:
Does modern physics use a different kind of logic or thought process from that which is used in other sciences?

Copyright©2018
IolziB
by Douglas Reinhardt
All Rights Reserved

QUANTUM QUANDARIES AND QUACKERY

Abstract

This book is a continuation in the series entitled THE PHILOSOPHY OF SCIENCE APPLIED TO MODERN PHYSICS. Although Quantum physics is a technical field, it can be critiqued in terms of the assumptions, inferences and the philosophy of science used in building theory. While the picture of the atom works very well, the conclusions drawn from double-slit experiments lead to contradictory interpretations as to what is going on at the invisible atomic and subatomic levels regarding waves and particles. Despite the conceptual contradictions, the formal math of each interpretation is said to equally valid and accords well with experimental results. Some of the contradictions include: the wave collapses to a particle in the Copenhagen interpretation vs. the wave doesn't collapse in the Many Worlds theory, the particle doesn't have a definite state until observed in Copenhagen vs. the particle has a definite state from inception in Bohmian Mechanics. Much of the theory of Quantum Mechanics has been largely influenced by Eastern mysticism, and its founders, including Bohr, Heisenberg, Bohm and Capra, were mentored by Eastern mystics of one tradition or another. The influence of Eastern mysticism is apparent in Quantum mysticism which some physicists deny, and others embrace with the passion of devotees. Thus, much of Quantum theory is unscientific and violates the principles of empiricism, scientific realism and objective, mind-independent reality. Entanglement theory and experiments are examined from the point of view of scientific, critical thinking. There is much vagueness and ambiguity in entanglement theory allowing physicists much slipperiness in denying obvious contradictions. For example, some physicists claim that if the experimenter forces a particle into a particular state, entanglement is destroyed and communication between particles is lost. Of course, the only way to test for non-local communication between entangled particles is to change one particle to see if the other changes accordingly. The hedge for this contradiction is that the quantum rule allows one such change to the particle to determine if the other particle responds in a complementary way. After that one change, entanglement is destroyed, and thus entanglement cannot be used as a system for instant communication at any distance. String Theory represents the culmination of the trend in Quantum physics toward mysticism and mathematical metaphysics. After decades of vain mathematical speculation on the behavior and shapes of strings, no empirical evidence has been found to support this decidedly unscientific theory. Overall, this book is an application of the scientific philosophy of rational-empiricism upon which the scientific method is based to critique the concepts of Quantum theory.

QUANTUM QUANDARIES AND QUACKERY

Dedication and Acknowledgements

I dedicate this controversial book by an author who has ventured far outside his field to my loving wife, Peggy, who has been so tolerant and understanding of my need to make a statement on my perception of the "state of science" particularly in the field of physics. In our post-retirement years when folks are supposed to be sitting in rocking chairs, traveling and socializing with friends and family, she has supported me in every way despite my taking time away from her to pursue this passion.

I am also indebted to my science-oriented friends who have continued to stimulate my interest in hard science over the years. Those who argued for the standard theories of physics and resisted my classical tendencies have made me sharpen and fine tune my arguments for an alternate way of viewing the "nature of nature" and bring the "physical" back into physics.

And, to all my family, who may have felt neglected because of my solitary pursuits in reading, thinking and writing, I appreciate your support and understanding.

Table of Contents - toc

CHAPTER 1: THE NATURE OF THE QUANTUM p. 6
The Birth of Quantum Theory p. 6
Later Versions of Quantum Theory p. 8
What to believe when the quantum world and the macro Classical World clash p. 9
The Wave-Particle Duality or the Wave-Particle Unity? p. 10
The Medium for Electromagnetic Waves p. 11
The Double-Slit Experiment and Quantum Theory p. 12
Wave-Particle Holism p. 14

CHAPTER 2: THE VARIOUS INTERPRETATIONS OF QUANTUM THEORY p. 17
Copenhagen Interpretation: Heisenberg's Uncertainty Principle p. 17
The Many-Worlds Theory p. 29
~The Quantum, 2-Bit Computer and the Many Worlds Interpretation

Bell's Theorem of Inequalities and Quantum Entanglement p. 37
~Procedure for Testing Bell's Theorem and Entanglement: Copenhagen, EPR and MWI
~Simplified Entanglement Experiment
~Strong Entanglement: Beam of Entangled Electrons deflected by Magnet
~Proposed Experiment to Test EPR and QM
~Strong Entanglement vs. Weak Entanglement
~The State of Confusion in Quantum Mechanics regarding Entanglement
~Quantum Theory vs. Relativity: Can information Travel Faster than Light?
~Does Bell-Aspect's Refutation of EPR Confirm that Consciousness Creates Reality as in the Copenhagen Interpretation

The Seamless Whole: de Broglie-Bohm Pilot Wave Theory p. 72
Zero World's theory and Information theory p. 75
Decoherence and Wave Collapse p. 75
QBism: Demystifying Quantum Weirdness? p. 77

CHAPTER 3: MORE WEIRDNESS IN QUANTUM THEORY p. 80
Consciousness, Measurement and Wave Collapse p. 80
Quantum Tunneling p. 80
Delayed Choice and Quantum Eraser – Mystical, Metaphysical and Physical Interpretations p. 81
Wave as Information or Wave as Physical Reality p. 84
Technology and QM p. 84

CHAPTER 4: QUANTIZING THE UNQUANTIZABLE p. 86
String Theory – the Ultimate Quantum Theory p. 86
Quantizing Space, Time, Spacetime and Gravity (Loop Quantum Gravity) p. 90
Light as a Waving Stream or Streaming Wave p. 92
Interaction of Quantum and Macro Worlds p. 95

CHAPTER 5: THE STANDARD MODEL OF PARTICLE PHYSICS p. 100
Spin Direction and Magnetism p. 100
Electricity and Magnetism related to Electro-chemical Processes p. 103
Gravity, the weak link in the quantum chain p. 104
Supersymmetry (SUSY) p. 105

CHAPTER 6: UNIFICATION THEORIES p. 108

QED (Quantum Electro Dynamics): The Normalization of Renormalization p. 108
Electroweak Unification p. 110
CHAPTER 7: CONCLUSIONS REGARDING QUANTUM PHYSICS p. 112

References p. 114

CHAPTER 1: THE NATURE OF THE QUANTUM

Einstein: God doesn't play dice.
Darwin: Nature plays dice.
Intelligent Design: God or nature plays with loaded dice.

New Conservation Law: Quantum Mechanics is the conservation of weirdness.

Einstein said: "The most incomprehensible thing about the universe is that it is comprehensible." However, Feynman said that he could safely say that no one understands the quantum – the stuff the universe is made of.

To the uninitiated, quantum mechanics seems like magic. Like the magician who can make a white rabbit disappear and reappear somewhere else (as though it were neither here nor there until the magician makes it appear and have a definite place in time and space), so nature gives us the sleight of hand with subatomic particles according to quantum theory. However, we know the magician has only fooled our eyes, but our minds tell us that s/he has not really violated any of the classical laws of nature. But, the quantum magician tells us that it is not that nature has tricked our senses and minds, but that subatomic particles actually behave in a magical way that does indeed violate the classical laws of nature that govern massive objects. This book, however, questions whether nature at the micro level has given us the sleight of hand or whether our instruments are still not sensitive enough to see through nature's magic tricks or whether the laws of nature are really different at the level of the very small. Einstein, although he advanced his own counter-intuitive theories in relativity, believed that much of nature at the micro level was hidden from our view and too subtle for our instruments and theories to see. He believed that physicists were on the outside looking in and trying to figure the internal workings of a system hidden from us. It is rather like trying to figure out what makes a watch tick when you cannot see inside the watch.

The Birth of Quantum Theory

Quantum theory was born as a result of classical theory's failure to explain black body radiation. Now a black body was originally conceived as a black box (hence the name black body) made of a dark material such as graphite which ideally absorbs all light or electro-magnetic radiation and emits all frequencies of radiation. Total absorption means that the box hypothetically does not reflect any light or radiation. The box is made with a small opening in the front into which radiation is projected. Since the radiation comes in at an angle, it is bounced around in the box with little chance of escaping back through the small opening. Thus, the radiation is eventually absorbed. As more radiation is poured into the box, it begins to heat up and at some point, it begins to glow as the radiation reaches the visible part of the electro-magnetic spectrum. As it continues to heat, the colors (frequencies) of light change through the colors of the rainbow, first beginning as red, then yellow…and ultimately violet light (Clegg 2014: p.16-18). Thus, the changing color or frequencies indicate a higher energy level. The mathematics of classical theory (formulated by Rayleigh and Jeans who visualized light as a wave) predicted an "ultraviolet catastrophe". The ultraviolet catastrophe means that the radiation energy emitted by the black body becomes infinite when the black box is heated beyond visible light, i.e., beginning in the ultraviolet frequency. Of course, if it were true, the ultraviolet catastrophe would be a catastrophe not only for physics theory but for the universe as well which would be destroyed by fire, rather than ice. This is a classic case of the mathematics not corresponding to physical reality.

Since the observation did not match the math, Planck came to the rescue. He devised a constant to plug into the equation, which is a very small number, which indicates that light is not a continuous wave, but is comprised of small, discrete units which he called quanta, and was later named photons (photo-electrons). Allegedly, Planck thought that what he had discovered was a mathematical trick that made the equations conform to observation. He had no theoretical explanation for the constant except that it indicated that light was made of packets or particles.

Now, the question becomes how does the fact that light increases in frequencies by quantum leaps rather than continuously mean that energy does not continue to increase as the frequency increases and wavelength shortens. Why does the energy level go down at some point after passing the ultraviolet frequency and reverse the trend of "the higher the frequency, the higher the energy?" Einstein provided the theoretical answer to the puzzle and it seems rather circular. The amount of energy in the box is finite, so the energy cannot continue to go up to infinity even if the frequency of the radiation continues to rise. Perhaps, Einstein's work on the photoelectric effect provides a clue as to why the black body radiation is a curvilinear relationship rather than a straight-line relationship. Einstein found that if metal is hit with light of higher frequencies, electrons are knocked off the metal. No electrons are knocked off until light is ramped up to the yellow frequency and more electrons are knocked off as the frequency goes up to the violet and ultraviolet levels. Then the only way to knock off more electrons is to increase the intensity of the light, that is the volume or amount of light, but there is a limit as to how much light of any frequency can be generated.

As indicated here, Einstein was in agreement with the beginnings of quantum theory which he helped to pioneer. Obviously, he believed in the quantum because the theory corresponds to observation and there is a rational theoretical explanation for it. However, as quantum theory evolved, particularly with the double-slit experiments which allegedly showed that light was a particle and a wave and that both faces of light could not be shown at once, he became quantum theory's greatest critic. Einstein could not buy these mystical and non-rational theories that were given the more polite designation as counterintuitive. I am in agreement with Einstein on this issue even though I also disagree with what I perceived to be the counterintuitive aspects of relativity.

Later, Bohr would explain the atom in quantum terms. Bohr, at first, envisioned the atom as a miniature solar system in which the nucleus was like the sun while the electrons were like the planets orbiting around the center. However, unlike the solar system, electrons could not orbit in just any space around the atom – they had to orbit in fixed paths around the nucleus – and they could not travel in-between these fixed orbits. When heat is added to the atom, an electron would be booted up to a higher orbit or higher energy level (but not in-between). Since these fixed orbits were definite energy levels or steps, they represent quanta, and an electron has to make a quantum leap from one level to another. When the atom lost energy, the electron would drop to a lower level and give off light because of the energy it had acquired when being booted to a higher orbit. The greater the energy imparted to the atom, the higher the quantum leap and when the electron jumped to a lower level, the energy level would be higher when making the largest jumps. A big jump emits higher frequency light (higher energy) and a smaller jump emits lower frequency light (less energetic). To visualize this, imagine that energy is like an inclined plane so that energy is continuous and can occupy any place along the plane. On the other hand, imagine a set of steps where energy can only occupy certain discrete levels. Perhaps these quantum steps would be what Planck discovered mathematically as a discrete unit of energy. However, this idea of electrons having definite places on steps seems to contradict Bohr's idea that an electron occupies no definite place until it is measured or observed. Bohr might argue that the electron is indeed in a fuzzy, indefinite location until a measurement is made. However, this argument flies in the face of the reality that atoms and electrons were doing their thing, jumping levels and creating light long before humans arrived on the scene.

> The colors of the rainbow indicate the quantum levels of light. Each color is distinctly separated from the other rather than being continuous and blended. Each color represents a different frequency and thus a different energy level. To create secondary colors, various frequencies can be converged to create an interference pattern that either adds or subtracts from one color frequency or another. Complementary colors actually cancel each other to a gray in paints (reflected light) because the trough of one meets the crest of the other in destructive interference. However, in non-reflected light, complementary colors produce white light. On the other hand, adjacent colors may create constructive interference or harmonies as in sound. The angle that the light waves intersect may determine the specific hue of a secondary or tertiary color.

Later Versions of Quantum Theory

Is it possible for a theory to be any more counter-intuitive than relativity with its emphasis on spacetime (an abstraction) as a real physical force? Einstein didn't think relativity was metaphysical or mystical. However, even though he helped to pioneer the field, he never accepted later quantum mechanics as a complete theory, insisting that it was only a partial picture of reality – partial because of hidden variables. The word "quantum" means "discrete quantity" which means that energy does not come in a continuous stream as we might perceive it, but rather it comes in discrete packets called quanta as defined by Planck's constant. Hence, light comes in quanta called photons rather than being a continuous stream of energy. Quantum Mechanics deals with the behavior of energy and matter at the atomic and subatomic levels. We are told, right out of the gate, that particles of this size do not behave the same way as massive objects as described by classical physics. Einstein had already told us that massive objects do not behave the same way at high speeds approaching light, and now we are told that micro-particles of light (photons) do not behave like anything else known to physics.

Terms used in quantum physics that are different from their standard meaning

I thought it might be useful at this junction to remind the reader of some terms that are used in physics that are different from their standard, dictionary definition. In Modern Physics, there is much confusion about the meaning of nothing, vacuum, zero and negative. These definitions might help to clarify.

1) **Vacuum** means there is no matter including gas in a given volume, but radiation in the form of photons vibrating at various frequencies is present in the vacuum. Thus light, electro-magnetism and neutrinos are widespread in the universe – even in areas called "the vacuum". Of course, the fallacy of this concept is that matter and energy are treated as separate entities when the equivalence and interchangeability of matter and energy is a well-accepted dictum in physics.

2) **Zero-point energy** – zero energy does not mean **no** energy in physics. Zero-point energy is the minimum known energy or radiation in a given volume of space. Perhaps the void in intergalactic space would contain low levels of radiation and that would be the lowest level of energy in the universe, but it would not be zero energy in the literal sense of zero as "nothing" since space does have heat as indicated by the cosmic microwave background radiation.

3) **Virtual photons** popping up out of the vacuum or void and disappearing – I would say the photons were already there from radiation spewed into space by trillions of stars. There is a surge of the radiation at certain times and places – much like a surge of atmosphere that we call wind. The air was there before it started moving, and the photons were there before they surged and manifested themselves – they did not pop up out of nothing as physicists often say. But then, again, nothing is not really nothing in physics semantics.

What to believe when the quantum world and the macro Classical World clash?

When observations and measurements are in conflict between the quantum world and the macro classical world, which are you going to believe? Here are the options as I see them.

1) The quantum world is real, and the macro world is illusion. This seems to be the consensus view of modern physics. N. David Mermin espouses this view: "We now know that the moon is demonstrably not there when nobody looks (Herbert 1985)." This idea is derived from John Wheeler's participatory anthropic principle based on his interpretation of the delayed choice experiments in quantum physics. Other physicists believe that the visible universe is not made of solid matter but is some kind of holographic projection. In John von Neumann's famous book *The Mathematical Foundations of Quantum Mechanics,* he offers a proof that if quantum theory is correct, the world cannot be made of ordinary objects. Paul Davies (2006) says:

Things may be peculiar down among the atoms, but they are so small, who cares...because to larger items such as tables and chairs, these effects seem insignificant. But you can't get away with that (kind of thinking) because, after all, tables and chairs are actually made of atoms – they are quantum objects...We and everything else are quantum systems...Even though these (quantum) effects are tiny, they are not insignificant (quote from *The Anthropic Principle* video).

2) The macro world is real, and the invisible quantum world is not yet fully understood. Quantum weirdness often arises because we cannot observe and measure atoms and subatomic particles the same way we make measurements in the macro world. Einstein, Schrodinger, deBroglie and even Bohm seemed to espouse the view that the quantum world would be much more similar to the classical world if we could observe it directly. Perhaps uncertainty is due to our ignorance rather than the unpredictability of the quantum realm. One test of the reality of quantum theory involves logical extension of the theory into the macro world. If logical implications and predictions from the quantum world to the macro world lead to a *reductio ad* *absurdum*, the quantum theory is probably wrong. For example, if quantum theory projects that cats can be dead and alive at the same time, human observation created the universe and its history, and quantum superpositions (probabilities) result in the universe splitting myriad times so that each probability becomes a reality in some universe somewhere, I would wager that there is something wrong with the original theory that predicts these outlandish extensions into the macro-classical world.

3) Both worlds are real. The quantum laws rule at the micro level and classical laws rule at the macro level, and somehow the quantum uncertainties average out to create the certainties of the macro world. Some physicists have invoked the concept of decoherence to explain how the atomic world with all its weirdness becomes the classical macro world that we have evolved to understand. Here we might invoke the principle of the holistic principle, i.e., the whole is greater than the sum of the parts, or, more precisely, the whole is different from the individual parts that make it up. As an example, the chemical properties of water are very different from the chemical properties of hydrogen and oxygen taken separately as elements.

An excellent example of this disconnect between the quantum and macro reality is Schrodinger's parable of "dead-and-alive" cat. Obviously, the idea of a simultaneously dead-and-alive cat violates rationalism and empiricism since no such animal could be observed or understood logically. So, which are we to believe – the micro-world where allegedly things can exist and non-exist or the macro world where things either exist OR non-exist? Classical science is based on empiricism. Since a dead-and-alive cat cannot be observed directly, nor can any indirect evidence support such a state in an animal, then at the macro level, we would have to say that such a prediction would be unscientific. More on this conundrum later. The table below illustrates these three views.

	Micro Quantum World	Macro-Classical World
Classical rules all	X	X
Quantum rules all	X	X
Quantum rules micro only	X	
Classical rules macro only		X

The Wave-Particle Duality or the Wave-Particle Unity?

Before launching into a dissertation of waves and particles, it is always advisable to define our terms and explore the relationship between these manifestations of nature. I take particles to be pieces of matter that make up massive objects, and I see waves as being the energy associated with all mass. Einstein's $E=MC^2$ tells us that mass and energy are interchangeable and thus particles and waves are interchangeable. De Broglie tells us that even massive objects have waves associated with them. In this view, then, the relationship between particles and waves is that they are manifestations of the same thing that we might call mass. Einstein informed us that even energy has mass. Hence particles would be condensed energy bound up in matter, and energy would be rarified, high speed mass spread out. The same principle applies to macro matter. When matter is heated and energized, it expands; when cooled and less energetic, it condenses. Thus, particles and waves are part of the same phenomenon, but they manifest themselves differently depending on their energy state. Hence, the two faces of mass can only show themselves in one state or the other, and probably not in an intermediate state since matter-energy seems to exist as quanta. This notion of dualism is supported by the fact that in the double slit experiment, photons and electrons show themselves either as a wave or particle but never as both.

The other view of the particle-wave relationship is that the particle and wave exist together, and the particle is somehow embedded in the wave. Since waves seem to need a medium (just as sound waves and water waves do), it seems reasonable that the wave is the carrier of the particle in the vacuum of space or perhaps the particles provide the medium for the waves to be propagated. Hence physicists speak of locating the particle somewhere in the wave and indicate that only the probable position can be known until the wave collapses and reveals one definite position of the particle. Hence, in this scenario, when the wave collapses, the particle can travel by itself since the wave is no longer present. The fact that light can travel through the relative emptiness of space indicates that photons are the medium in which light waves are carried. Thus, in this view, the wave and the particle travel together, but for the particle to manifest itself, the wave carrying it must collapse. However, it is uncertain as to whether the wave is carrying the particle or if the particle is carrying the wave.

A third way of looking at the wave-particle relationship is that the particle and wave are bound together and that the wave never collapses – the experimenter just forces the photon to show one face or the other. In one scenario, the wave doesn't collapse but it loses information and becomes decoherent. In Bohmian mechanics, however, the wave does not collapse, nor does it lose information and become decoherent.

	Table: Wave-Particle Unity	
	MASS	
	Energy	Matter
Particle		X
Wave	X	

There is much confusion about the relationship between the particle and the wave. There are at least two conceptions of the relationship.

1) The particle spreads out as a wave and becomes massless; then condenses as a wave into a particle and regains mass.
2) The particle is a carrier of the wave; therefore quantum physicists try to locate the position of the particle, but finds that the position of the particle and its momentum can't be known at the same time.
3) The wave is the carrier and guide of the particle.

It would seem to this writer that the wave and particle have to exist simultaneously because the wave needs particle to carry it even as water waves need molecules of H_2O, and sound waves need air or other medium of particles to carry its pressure wave.

QUANTUM QUANDARIES AND QUACKERY

Nature of the Quantum

The Medium for Electromagnetic Waves

Since water waves and sound waves require a medium of particles in which to be propagated, it would seem reasonable that electromagnetic waves would also require a medium. Early physicists assumed that space was made of some sublime matter which they called *ether* which carried light and electromagnetic waves. Even Maxwell believed that space was filled with ether which carries electromagnetism. However, Michelson and Morley's experiment could find no evidence for this rarified medium called ether and this provided the basis for Einstein's Special Relativity which requires space to be a vacuum (no ether) in order for light speed to be the ultimate constant. Later, quantum physicists hypothesized that space is filled with a quantum foam of virtual photons popping in and out of existence. Yours truly believes that ether is not necessary to carry light waves because light is a particle and a wave, and the particle, the photon, is the carrier of the wave. Follow this line of logic if you will:

Waves Need a Medium of Particles to Wave in.
1) Water is a pre-existing particle medium for water waves and the water can be still or moving.
2) Air (or other medium) is a pre-existing particle medium for carrying sound waves of various frequencies and the air can be still or moving.
3) Photons are a particle medium for carrying light waves which travels with light and there is no need for a pre-existing medium. The photons travel with the wave and are not static. Hence the light wave is more like a waving stream in which particles move with the wave rather than a still lake where disturbances in the water make waves on an otherwise placid surface. The table below illustrates this concept that light is a moving particle that carries the waves.

	Types of Waves		
	Water Wave	**Sound Wave**	**Light Wave**
Particle Medium	Water molecules	Air Molecules	Photons
Wave	Rhythmic disturbance	Rhythmic vibration	Vibration of photons

The Double-Slit Experiment and Quantum Theory

The essentials of quantum theory can be demonstrated by the double-slit experiment conceived and pioneered by Young in 1801. The following are steps involved in the experiment.

1) Particles (photons, electrons, etc.) are beamed through a single slit and reveal themselves to be particles.
2) Then particles are beamed through a double slit apparatus and show themselves to be waves because an interference pattern emerges on screen (Interference and diffraction are opposites, interference involves combining waves, diffraction involves separating waves based on different frequencies). Interference patterns indicate that the particles behave as a wave.
3) The wave pattern is shown in constructive and destructive interference and indicates the wave was spreading out after it goes through the double slit. But the impressions on the film or screen are **points** indicating that there were point particles carried by waves.

4) In route to the slit and screen, the position, momentum and path of the particles are said to be uncertain.
5) After passing through a slit in route to the screen, a particle may be measured to determine which slit it went through.
6) As a result of the measurement, the wave collapses and a random pattern of dots emerge on the screen manifesting particles.
7) However, since the particle was interfered with in some way, it was made to manifest as a particle.
8) Therefore, the particles position and velocity cannot be known at the same time since you have to change its velocity to find its location. So far, there doesn't seem to be a mystery.
9) But, since the path of the particle on the way to the screen is unknown, the particle could be anywhere in the universe (non-local).
10) Since the particle's position is not identifiable while also knowing its velocity, the particle itself does not have a definite position, or it occupies all possible positions on its way to the screen. The particle only takes on a position when the experimenter observes it or makes a measurement and thus collapses the wave. Theoretically, this anomaly is not due to the experimenter's ignorance of where the particle is - it is a fact of nature that the particle does not have a location until observed and brought into existence. The following table shows the stages of the process of the double slit experiment and where the uncertainty resides.

Input ==>	Interaction of particles ==>	Output
Particles shot toward slits	Particles approach and pass through slits and from slits to screen	Particles recorded on photographic screen
Certainty as to what particles are emitted	Uncertainty as to state of particles in route. Thus, the particle is thought to have no certain location or is in two or more locations at once.	Certainty about mathematical formalism which predicts the probability of finding a particle at any given location.

The above table shows where the uncertainty and mysticism arise regarding quantum mechanics – it is in the middle stage where particles leave the particle gun and hit the photographic screen. It is in this stage that particles are said to have no specific status – it is as if they do not exist until their behavior is observed on the photographic screen. If the experimenter identifies the position of the particle, then she cannot know the velocity of the particle because she has changed the velocity in locating it. In contradiction, theory says that a particle can have two contradictory statuses at once (the particle can be in two or more states at once – yet can't have position and velocity at the same time). The middle stage is what Einstein says is the watch that is un-openable. One can only observe what one does to the watch (winding and resetting time) and can see the output in terms of telling time, but one cannot see inside the watch to know what is making it tick. It is in this middle stage that several mystical interpretations have arisen to explain what goes on with the particles in the in-between zone. These interpretations range from the Copenhagen to the Many Worlds, Pilot Wave, etc. However, despite all this uncertainty about the middle zone, quantum physicists can argue that quantum mechanics is a precise science because, although they cannot predict the behavior of any given particle, they can predict a statistical pattern of where particles will land on the screen in accordance with Schrodinger and Heisenberg's formulas and tables. However, some theorists, such as Deutsch, argue that their interpretation of the middle zone has the status of a sound theory rather than an interpretation of what goes on.

Double Slit Mystery – Possible Resolution
When a single electron is shot with both slits open in the double slit experiment, a single electron is said to be in several places at once, take all possible routes to the screen and therefore pass through both slits and over time create an interference pattern on the screen.
1>The idea of emitting one electron or one photon at a time is a fallacy I believe. The size of an electron is estimated to be 10^{-16} meters, so the notion of shooting only one of these invisible particles at a screen is implausible at best. What is probably shot is a clump of these particles which makes a visible dot on the screen. One electron would probably not make a dot large enough to be seen on a photosensitive screen.
2>Unless the experiment is performed in a vacuum, a single electron would interact with particles (including other electrons) in the air. This could create a pattern of interference on the screen.
3>If the quantum foam is real and there are particles constantly popping in an out of existence, these particles could also interact with electrons or photons shot toward the screen and create interference patterns.

Wave-particle holism

The following are ways that show that the wave-particle duality is a false dichotomy. Instead, we should refer to this phenomenon as wave particle unity.
1) The wave and particle are inextricably linked. There is no collapse of the wave in the sense that the wave ceases to exist. The wave only appears to collapse and can be brought back to observation with a polarizer. De Broglie found that there is also a wave associated with macro objects.
2) When the experimenter puts a detector at one slit to determine which slit the particle went through, the detector interferes with the particle and makes the wave appear to collapse.
3) As long as the experimenter is shooting particles through a single slit, the particles travel in straight lines and their waves do not interfere with each other. That does not mean that there were no waves present in association with the particles.
4) I believe it is an illusion that the experimenter can shoot only one photon or electron at a time and if both slits are open, it goes through both slits and interferes with itself. Consider the fact that the photon is a hypothetical particle that is said to be massless and which is so small that no microscope is able to see one, and although the electron is larger than the photon, it cannot be seen either. Hence, it is highly doubtful that the experimenter is emitting only one photon or electron at a time. In all probability, the experimenter is emitting a small stream of photons or electrons even at the lowest level so that there are enough particles that their waves will interfere with each other in going through the double slits. Furthermore, even if the experimenter could shoot only one particle through the slits, the particle would interact with similar particles unless

the test chamber is evacuated. We know that light slows down to 75% of its vacuum speed in atmosphere and scatters. Even, if the chamber is evacuated of air, the Casimir effect indicates that there are subatomic particles that manifest themselves in a vacuum. A vacuum is defined as the absence of atmospheric gases, but we know there is no perfect vacuum. Thus, we see that there are many uncontrolled variables in these double-slit experiments.

Furthermore, experiments with water waves show the same kind of interference patterns as subatomic particles, so that it does not appear that subatomic particles behave any differently from supra-atomic particles such as molecules of H_2O in liquid form, thus indicating that the extent of quantum weirdness has been greatly exaggerated. If the experimenter could shoot one molecule of water (which would be hard to isolate) through the slits, it would probably behave in the same way as photons and electrons in appearing to interfere with itself. Furthermore, de Broglie matter waves indicate that massive objects behave much the same way as micro-particles.

> In the double slit experiment, particles always manifest themselves as particles – they do not manifest themselves as continuous waves. If the particles are allowed to go through only one slit or if they are interfered with via measurement, they manifest as a random scattering of dots. If on the other hand, the particles are allowed to go through both slits and are not measured, they arrange themselves as a dot pattern than appears as a wave with an interference pattern. This pattern becomes clearer over time after many trials and many dots on the screen.
>
> Another way to state the Heisenberg principle is that you can't know the location of the particle and have interference at the same time.

The following table is a compilation of the various interpretations of the wave-particle duality largely based on the double slit experiment and other observations.

QM Theory	Determinate vs. Indeterminate	Wave Collapse	Hidden Variables	Local vs. Non-local	Consciousness Effects	Faster than Light
Copenhagen	Indetermminate	Yes	No	Non-local	Yes	Yes
EPR (Einstein, et al)	Determinate	No	Yes	Local	No	No
Many Worlds	Determinate	No	No	Non-Local	No?	Yes
Bohm's Holism	Determinate	No	Yes	Non-Local	Yes	Yes
Bell's Theorem	Indeterminate	Possibly	Yes	Non-Local	No	Yes
Zero World	Determinate	No	No	Non-local	No	Yes

According to Sean Carrol in his video *Quantum Mechanics (an embarrassment) - Sixty Symbols*, here are the preferences among physicists for various interpretations of quantum mechanics according to a recent survey:

Copenhagen 42%
Many Worlds 18%
Information based 24%
Objective collapse 9%
Quantum Bayesianism 6%
Relational QM 6%
Other 12%
No Preference 12 %
Total =129%

Now I am not going to make a big issue out of the fact that there are 129% physicists in 100% of the physics community, but this error seems rather similar to some of the thinking about quantum phenomena in which a cat can be dead and alive or a particle can be here and there at the same time. As indicated by quantum physicists, normal logic does not apply. While some physicists claim that all these interpretations are equally reliable in predicting quantum phenomena, that assertion is hard to believe when there are some many divergent interpretations. Perhaps by observing the statistical dot pattern that emerges from a double slit experiment, predictions of the results of future experiments with the double slit can be made with statistically significant accuracy. However, to say what happens between shooting the particle and its detection on the screen behind the slits is up for grabs. Now, I have read that more physicists are deserting the Copenhagen interpretation (because of quantum mysticism that appears to support all manner of mystic philosophies), and, like Sean Carroll, are moving toward the Many Worlds Interpretation of Everett. If one is looking for a more rational, less mystical theory, then the Many Worlds Interpretation seems an ironic choice indeed. A theoretical interpretation that indicates that the universe splits to form new universes for every quantum superposition that exists but that these other universes can never be observed, is hardly more rational than the Copenhagen mysticism. The following discussion underlines what this author sees as the great confusion that abounds in quantum theory.

CHAPTER 2: THE VARIOUS INTERPRETATIONS OF QUANTUM THEORY

Copenhagen Interpretation

I cannot believe that the Moon exists only because a mouse looks at it. (Albert Einstein)

Neils Bohr could not be described as a pure empiricist or a logical positivist. He believed that the metaphysical could not be separated from the physical, the subjective from the objective, and the experimenter from the experiment. In other words, the mind is entangled with physical reality as one system.

> **Copenhagen Mysticism**
>
> **Skeptic:** If a tree falls in a quantum forest and no one is there to observe it, did it actually fall?
> **Copenhagen:** No, the tree has not fallen because there is no observer there to actualize the probability of its falling and therefore make it an accomplished fact. The tree is in a state of superposition in which it is both fallen and standing until someone observes it and makes it one way or the other.
> **Skeptic:** But if someone comes along later and sees the tree has fallen, does that not prove that the tree had fallen before?
> **Copenhagen:** No, it is not fallen until an observation is made of the fallen tree, then the observer's consciousness goes back in time and actualizes the tree's tendency to fall and creates the history of the fallen tree up to that point. John Wheeler has demonstrated this with delayed choice experiments.
> **Skeptic:** So, cause does not always precede effect in time; effect can come before cause.
> **Copenhagen:** Yes, according to Wheeler that is the way the universe was created by human observation going backward in time.

The above imaginary conversation illustrates what has become known as quantum mysticism. While some physicists deny the connection of quantum physics to Eastern mysticism, others have actively embraced it. The work of Fritjof Capra, who wrote the *Tao of Physics* and Gary Zukav, who wrote about *The Dancing Wu Li Masters,* details this intimate relationship between physics and mysticism. Some physicists apparently think that papering over quantum mysticism with elaborate mathematics makes it scientific. This author believes that the mysticism of quantum physics is caused by the mystery of atomic reality which cannot be directly observed. Mysteries presented to the human mind almost inevitably lead to mysticism. Hedges like those provided by Wheeler are an escape hatch from falsification.

Heisenberg's uncertainty principle indicates that it is impossible to know the position and velocity of a particle at the same time, not because of any measurement problem by the observer, but because nature at the quantum level is random and unpredictable, so the information is not there for the observer to measure. The particle therefore does not have a definite position, path or speed until it is observed or measured – it only has a probability of being in one state or the other, or perhaps it is in all states until a measurement is made. Thus, the observer's consciousness, not the obtrusiveness of the method, causes the wave to collapse and causes one of the probability waves to become a physical reality. Before that collapse, the position, momentum and path of the wave is not only uncertain, but does not exist or exists in all possible states. It's not that you are finding the particles position; you, the observer, are creating its position and thus constructing reality. Allow me to critique this interpretation point by point.

Heisenberg's uncertainty principle is that there are pairs of complementarities that cannot be known at the same time, for example, the position and momentum (speed) of a particle cannot be measured at the same time. According to Heisenberg, this is not a failure of measurement, but uncertainly is inherent in nature. Likewise, Bohr believed that nature cannot be known directly but that knowledge is always a mixture of nature and the mind interacting with each other. So, perhaps, it should be said that these complementarities are inherent in mind interacting with nature, not in nature itself. Of course, the Copenhagen interpretation would say that mind is a part of nature.

Assuming that there is an objective reality outside the mind, the question that arises is whether the particle has a location and speed at the same time in the objective world. Speed is measured as distance/time. To know the distance, one has to know the location of an object in at least two different points in space, and to know the speed, one must know the location at two points in space and the time the particle took to travel between those two points. So, logic dictates that to know the speed, one had to know the position or location at two points. Of course, if a particle is moving, its position is constantly changing. How does Heisenberg answer this conundrum that one cannot know that speed without knowing the location? It is because the particle does not have a definite location or speed until a measurement is made. Again, location may be uncertain in the mind-nature combination, and the Copenhagen interpretation can make no claim on whether it has a certain location or not in the objective world. It comes down to the assumptions one makes about the nature of reality – epistemology, if you will. Science has been traditionally predicated upon the assumption that there is a mind-independent reality and that our senses and mind can closely approximate it. If our senses and minds did not approximate objective reality, then Darwinian principles indicate that we wouldn't have survived and evolved. Humans had to adapt to reality; reality did not adapt to them – reality of the physical world is not that fluid and forgiving. Whatever assumption one makes about reality, it does not remove the problem that to measure the speed, it is necessary to know (or closely approximate) the location of the particle in at least two points in space.

A corollary of the uncertainty principle is that one cannot detect a wave and a particle at the same time. If one takes a measurement when a quantum phenomenon is behaving like a wave, it instantly begins acting like a particle. Some theories indicate that the wave never collapses and that the wave and the particle are always found together. In the view of this author, many of these issues are a matter of semantics and linguistic inaccuracy.

Illustration of the Uncertainty Principle with Photography

Since I am a photographer of sorts, I thought of a way to illustrate the Uncertainty Principle even though, as a photographer, I am dealing with macro objects rather than micro, quantum objects. However, my viewing of macro objects is made possible by a quantum phenomenon – light. Let's say that I am photographing a car race. To locate the exact location/position of a speeding race car at any one moment in time, I would use a telephoto lens and zoom in tight to photograph the car with a mile marker in the background. When taking a single snapshot of the car in that position, I would not be able to tell the speed the car was traveling since I would need to know its location at two points in space and two points in time. However, I have its position at only one point in time and one point in space. Now to determine the speed, I might use a wide angle lens and take shots of the car in at least two points in space and time. Thus, by knowing the time and the distance, I would know the speed, but I would not have pinpointed the location of the car as precisely as I did with the telephoto lens zoomed in tight. However, there would be no reason I could not have shot the car with the telephoto at two points in space (rather than taking a since snapshot) and thus have known the exact location and exact speed of the car. Or, I could have used a video camera (which gives a very fast series of snapshots or frames) and calculated the position and speed of the car all around the track.

The difference between my photographing a macro object and filming a micro object is that I can see the macro object continuously and really know its speed and position at any time. However, in the double slit experiment, one can know the location of the particle at the beginning and end but cannot know exactly where it is in-between. Since this middle information cannot be known, one assumption in the Copenhagen interpretation is that the particle does not exist and therefore does not have a location/speed in space or time. Another assumption is that the particle is in superposition and occupies all possible positions in space and time between the beginning and the end, and only when a measurement is made in the end (on the back screen), does it take on one location in time and space. Another difference between by photographing a race car and photographing a quantum particle is that I am receiving reflected light from the car. However, in the quantum experiment, I would be hitting the particle with light or colliding it with a photographic screen. In the race car instance, I would be using passive observation and not affecting the race car's location or speed. The race car would be traveling at the same speed on the track whether I was photographing if or not. In photographing a quantum particle, however, I would be engaging in participant observation or active observation. I am interacting with the particle to determine its location/speed, and my interaction is changing those properties of the particle. However, I would assume that the particle has a location/speed whether I mess with it or not and that particles had these properties eons before humans were around to observe and mess with them. As a matter of fact, the interaction of particles and the forming of organic compounds is what led to human evolution according to Darwinian evolution which most physicists that I have read claim to believe in. To fix this problem, of course, physicists have to make the outrageous assumption that human consciousness went back in time to create this whole history of evolution. This Participatory Anthropic Principle, as it is called, gives humans god-like powers, yet many physicists claim to be atheists.

Heisenberg's Thought Experiment Demonstrating the Uncertainty Principle

The following explanation of Heisenberg's thought experiment, which I am critiquing here, comes from Gary Zukav's work (1979 : p. 111-113).

1>To see a particle, we must illuminate it with a light wave whose amplitude is smaller than the particle; otherwise the light rays will bend around the particle if the wave amplitude is larger than the object.

2> Gamma rays have the shortest wavelengths and lowest amplitude of any electromagnetic energy and are short enough to be blocked by an electron so that a shadow could be cast behind the electron onto a screen.

3> However gamma rays also have the highest energy of any electromagnetic rays.
Since they have the highest energy, gamma rays knock the electron out of its path and change its speed or momentum, but at least we know where the electron *was* when it got hit with gamma rays and if you keep hitting it with gamma rays you can know where it was at any given time. Therefore, in this scenario, we can know where the particle is or was, but we have affected its speed, so we can't know what its speed was before hitting it with gamma rays.

4> Here is the illogical part of *uncertainty*: Since the location and speed of an electron cannot be known simultaneously, the unknown factor doesn't exist. Out of sight; out of mind; out of existence. And this is not due to hitting the electron with gamma rays and altering its speed; this is inherent in the quirky nature of the electron itself. What kind of mystical logic says that I have affected an object, but my effect on the object has no effect on its behavior? My effect on the particle is the only reality that I can know, so it is the only reality there is. If I can't know it or see it, it doesn't exist. This sounds like Piaget's children who are still in the pre-logical stage of development: "If I don't see a ball that I saw previously, it doesn't exist anymore." This is also similar to primal thinking that anthropologists have learned from pre-Industrial societies: "The sun dies when it sets and doesn't exist again until it is reborn the following morning." It is a reasonable inference that the sun doesn't go out of existence each night because I can't see it, and it is a reasonable inference that a particle shot from a gun and reaches a photographic screen on the other side of the lab did not go out of existence between these two points and come back into existence at the other end of the lab when it hit the screen.

5> Given this unknowability of the electron, then Newton's laws of motion do not apply since his laws depend on knowing the position of an object and its speed simultaneously. For example, the formula for momentum (p=mv or momentum (p) is equal to mass x velocity) would not work because you cannot know a particle's velocity and position at the same time. The mass of an electron is said to be: $9.10938356 \times 10^{-31}$ kilograms, but if you can't know its velocity, you can't know its momentum. Therefore, when you locate where the electron is, then you can't know its momentum at that point in space and time.

Again, to be repetitive, since you can't know these properties of the electron, these properties do not exist, and since you can't know the properties to plug into Newton's formulas, then micro matter behaves differently from macro matter and Newton does not apply at this level. How is it that the parts of matter do not behave like the whole? How is it that the atoms that make up a baseball do not behave like the baseball itself – just because you can't measure the momentum and location of a baseball atom, like you can the whole baseball? This is the part of the Copenhagen Interpretation that I find most illogical and mystical and which violates a sacrosanct principle of science – that there is an objective, mind-independent reality. Atoms and subatomic particles preceded humans in time by well over 13 billion years – if one believes in the Big Bang. Furthermore, humans began to learn about atoms only some 100 years ago. So, electrons knew how to behave themselves long before humans tried to figure out their behavior. And, I cannot accept Wheeler's hedge that we humans created the universe with our consciousness going back in time for 13.7 billion years. Such a preposterous idea endows humans with god-like powers. Of course, the other hedge is that you can't understand the behavior of the subatomic world with logic – you have to accept it on faith in the mathematics.

6>But is Newton really pushed out of quantum reality altogether. The speed of an electron has been calculated to be 2200 kilometers per second (less than 1% of the speed of light). So, if we know the mass of an electron and its velocity, then we can know its momentum which is mass times velocity. The question becomes: "How did scientists calculated the speed of an electron, if they couldn't know its location in at least two points in time and space (velocity = distance/time)?"

7>If an electron has mass, then it should conform to Newton's laws regarding massive objects. Let's take Newton's Third Law which is: "For every action, there is an opposite and equal reaction." Now if gamma rays knock an electron out of its path farther than visible light because it has more energy or force, that sounds a lot like it is obeying Newton's third law. However, modern physicists say the photon has no mass, but this cannot be true since energy and mass are interchangeable as in $E=MC^2$. So, a photon with less energy will **not** knock the electron out of its path the same amount as a photon with more energy. Sounds like Newton to me.

8>Additionally, even if the experimenter slows down the electron after it has been hit by a gamma ray, it still has momentum even though it has less momentum than it did before it was hit since it has less velocity. To calculate that momentum, the experimenter could hit the electron a second time to determine its position, and the distance it had traveled in that period of time would indicate its velocity.

9>I would imagine that the way an electron's velocity is determined is to shoot a beam of electrons at a screen with detector. Since one would know the distance between the electron gun and detector, one would know the location of the electron in two points in space. Additionally, one would know the location of the electron in two points in time, so the velocity could be calculated. And one could calculate the electron's location at each point in space and time between the gun and the screen, assuming that its velocity was constant (inertial).

10>It just appears to yours truly that Heisenberg painted himself in the corner with this principle, and maybe he didn't have the technology that we have today to calculate the electron's speed.

More Critique of the Copenhagen Interpretation

1) **Particle has no definite position (location) until observation is made:** Another way of stating this concept is that space and time are not realities in quantum physics. But what exactly is Bohr saying about space and time? Is he saying that space and time do not exist at the quantum level, or that space and time exists at this level, but that the location of a particle cannot be known exactly in space and time? The fallacy of this line of reasoning to my mind is that in order to collapse the wave and define the particle's location, the experimenter had to interfere with the particle in some way such as hitting it with light. So, it is not the observation, in some sort of psychic way, that enabled one to create the particle's position; one affects the particles position with the energy that is used to locate it. Furthermore, the fact that the particle has a definite position in space at all times is indicated by tracing the paths of particles in bubble and cloud chambers and on film where the particle's speed and position can be traced or knowing a particles position in at least three points in the double-slit experiment. However, if you say that running the particle through a bubble chamber, or other device, is, in fact, making a measurement thus giving the particle a definite place in space, then let's look at fission track dating methods used in geology. Fission-track dating is a radiometric technique based on analyses of tracks left by fissioned particles in certain uranium-bearing minerals and glasses (Naeser 1979). Therefore, since the tracks were made long before any observation was made of them, the particles emitted

from Uranium had a definite position and trajectory in the absence of any human measurement. However, physics theory is difficult to refute, because there is always some epicycle-type hedge that can be brought to rescue a theory in trouble. For this rescue, we could bring in John Wheeler and the participatory anthropic principle. Wheeler would probably say that the fission tracks did not exist until some geologist or archaeologist discovered them and that the scientist's consciousness went back in time and caused a wave to collapse and the tracks to be made. Such speculation involves propping up one shaky theory with another shaky theory and could hardly be called science. This is what Popper meant by insulating a theory from falsification, so that whatever outcome an experiment produces, there is a hedge in the theory to cover it – thus such a theory proves contradictory results. Can anyone take seriously a theory that says we humans created the universe with our consciousness before we were created? This is what comes from treating time as an elastic phenomenon that can be stretched to prove or disprove any point.

The idea that at the quantum level, space and time are not realities, or perhaps that they are irrelevant, was upsetting to Einstein whose theories of relativity made space and time active partners in physical interaction on a par with matter and energy. Therefore, at the Solvay conference he presented a thought experiment in which a light source and a clock are suspended with a spring scale. When a photon is emitted, the scale will register a loss of weight since, contrary to other physicists who say that photons are massless, the photon as a particle of light energy has mass since matter and energy are equivalent as in $E=MC^2$. The time would be measured when this happened thus showing that photons (quantum particles) exist in definite space and time. This thought experiment was disconcerting to Bohr who spent a restlessness night trying to unravel this Gordian knot for quantum mechanics. During the night, Bohr had a eureka moment and the next morning informed Einstein that he (Einstein) had forgotten his own theory of General Relativity in which gravity affects time, so that the exact moment of the loss of mass of the light source could not be known definitely since time is personal and depends on one's speed and the level of gravity in which one is immersed.

However, Einstein need not have been flummoxed by this argument; because, if one had a clock with radium dial which emits radiation, the photons would be in the same gravity field as the clock itself; therefore the time would be the same for both photons and clock in that context. Or if cesium clocks were used, the radiation emission would be part of the clock mechanism itself, so the time for the clock and the particle would be the same. However, the quantum instant communication seems to strike a fatal blow to relativity's claim that nothing can go faster than light. Ironically, this notion of instant communication confirms Einstein's contention in this particular case, i.e., that quantum interactions take place in definite space and time. To wit, if entangled particles communicate instantly across space, there had to be a definite time in which they communicated, and the communication was between particles separated in space since *local*, hidden factors cannot account for this phenomenon. So, the communicating particles, when measured, had a definite place in space and when they were entangled, they had a definite existence in space and time –because they were entangled simultaneously. Therefore, if entangled instant communication is true, then there is simultaneity – another taboo in relativity. So, quantum entanglement rescues Einstein in confirming space and time, but goes against him on the speed limit of light and simultaneity.

QUANTUM QUANDARIES AND QUACKERY   Various Interpretations of QM

> It is ironic that instant communication of entangled particles is used as evidence in quantum mechanics that space and time are not factors in the quantum realm. To the contrary, the particles were created and entangled in a definite place in time, they traveled away from each other in space over a period of time and communicated instantly as they were measured in two separate places in space and time.
>
> Furthermore, it is an irony that Feynman said that when matter/antimatter pairs are hatched, the matter particle goes forward in time and the antimatter particle goes backward in time. First, forward and backward in time has no meaning if time is not a factor in quantum mechanics. Second, if the particles are going in opposite directions in time, they would never meet and annihilate each other (convert each other to energy in the form of photons).

2) **The behavior of subatomic particles is uncertain and unpredictable**: Quantum physicists speak of a probability wave where a particle might be located. Therefore, particles, like people, have an element of chaos and unpredictability, and we can only predict a probability that certain behaviors will occur. If this is so, then how does the most accurate clock known to humans, the cesium clock, operate on quantum principles of uncertainty? We are told that cesium clocks deviate from perfect time by only 2 nanoseconds (2 billionths of a second) per day. How do these clockmakers get this level of accuracy out of the uncertainty principle which is that the behavior of subatomic particles is random and unpredictable? Now some sources say that the more predictable macro-world in which we live is based on quantum averaging. That is, the behavior of particles is not totally random, and the average behavior represents a pattern which is predictable – in other words, the dice are loaded. Thus, Newton's laws of motion are predictable on large scale objects. However, it would seem that a clock whose quantum ticks deviate only 2 billionths of a second per day would have to do better than averaging such ticks. Furthermore, there is the problem of electronic devices, such as computers, which operate with great precision in microcircuits. Could this level of precision be based on uncertainty and averaging? One might argue that Carbon 14 dating could be based on averaging since there is a time range given to the age of the artifact.

Furthermore, the paths and momentum of particles in a particle collider must be known with certainty in order to enable the particles to collide. Also, decay of accelerated muons, which is offered as evidence for Special Relativity, is said to be extremely precise and predictable. I suppose that one might argue that this randomness and uncertainty occurs only when there is a superposition offering several possibilities. However, the atomic decay, itself, involves a superposition in that there are two possibilities: either a particle will be emitted or not in a given interval of time. Even when one slit is closed so that light and electrons behave as particles, would there still be uncertainty regarding the position and momentum of the particle *a la* Heisenberg? The experimenter would certainly know the particle's beginning position, the slit it went through, and the end position on the screen. However, it is difficult to imagine a situation where there is not a superposition of quantum particles.

Moreover, if there is a pattern that emerges in particle behavior (as it does in double slit experiments after multiple streams of particles have been shot through the slits), then the behavior is not totally uncertain, random and non-deterministic. If the behavior were totally random, then there would be a scattergram with no pattern. This means that God (or nature) does play dice, but the dice are loaded so that the universe is part randomness and part order – it is in between. In playing cards, the shuffle may produce a random assortment, but in the hands of an intelligent player, the cards take on order, which the player directs toward the goal of winning.

3) **Wave collapse and the wave-particle duality**: The wave collapse is said to reveal the particle implying that the particle was a field of pure energy that condensed into a particle. However, the dominant interpretation seems to be that the particle was somehow embedded in the wave and being carried by it. In essence, the particle was hiding somewhere in this nebulous field. However, the wave does not disappear and can be brought back with a polarizing filter oriented at the correct angle (Garrett 2011). In the view of this author, the wave-particle duality is a false dichotomy. The phenomenon is both wave and particle at the same time and does not have multiple personality disorder – appearing as a wave in one situation and as a particle in another. Every other wave phenomenon (water and sound) involves a medium in which particles vibrate and bump into each other. Light does not need an extrinsic medium because it is its own medium providing an energy-wave field in which the particle is carried. If it were not so, light could not traverse the relative emptiness of space and travel millions of light years. Light should be conceived as a stream of particles embedded in a wave – a stream and a wave combined in one. Thus, we have a kind of unification so often sought in physics.

4) **Schrödinger's cat box** thought experiment was a parody on the Copenhagen interpretation which created a paradox for the observer effect. Schrödinger, who had heretical classical tendencies imagined that when a cat's life depends on random quantum event such a radioactive element emitting a particle whose probability wave is 50-50, the cat is both dead and alive until the observer opens the box, collapses the wave by making an observation and determines the state of the poor cat. After all, if a particle can be here and there at the same time or be a particle and a wave at the same time, then a cat can be dead and alive at the same time. Although Schrödinger meant this thought experiment to be a parody to reduce the Copenhagen interpretation to absurdity and thus as a defense of classical physics, some physicists continue to use this as an accurate description of the uncertainty principle of the Copenhagen interpretation. However, one can imagine placing a video camera in the cat box with a timer, so the state of the cat could be determined afterward without human observation. No problem to a quantum physicists like Wheeler, who would probably say that when a human looks at the video, his consciousness goes back in time and determines the moment when the video camera recorded the cat's death.

The Schrodinger cat box paradox is taken seriously (not as a parody) by Eugene Wigner, who created the thought experiment known as **Wigner's friend**. In this *gedanken*, Wigner performs the cat box experiment and leaves a friend in charge of the laboratory. Now the friend becomes part of the superposition along with the cat, so that when Wigner returns to the lab, his friend, who loves the cat, is both happy and sad about his pet's fate, and the cat is both dead and alive until Wigner makes an observation to collapse both waves rendering the cat either dead or alive and his friend either happy or sad (Wigner 1961). Of course, the ultimate state of the cat and the friend being either live cat/happy friend or dead cat/sad friend is a problem only in the single-world Copenhagen interpretation. In the many-world's theory, the cat and the friend remain in both states as the universe splits in two; so that both states are preserved – one state in one universe and the other in another universe. Believe it or not, mathematics have been worked out to show these probabilities.

Of course, the idea of the mind creating reality violates Popper's principle of a mind-independent reality which is the foundation of objectivity in classical science. How human beings, who are the latest arrivals on the planet, could have created everything that preceded them with their observation, is mysticism, not science, since science requires the cause to come before the effect in real time. Furthermore, how does one observe something that is already there, but not yet there, until he observes it? This is imagination, not observation. Why physicists would take offense at mystics and new agers latching onto this theory as proof of their beliefs can only be explained by sociology, not scientific considerations? It requires mental gymnastics to disentangle this theory from mysticism.

Moreover, some physicists do not take the cat-in-the-box thought experiment as a parody but actually accept as reality that the cat is both dead and alive until the observer makes a measurement. Yakir Aharonov, who was a student of Bohm, contends that there are two probability waves, not one, that determine the hapless cat's fate. He says that there is one wave coming from the past and another wave that is coming backward from the future, and when the two meet in the present, they collapse to become material reality. Thus, the poor cat is caught in a cross-fire between future and past (Discovery Science Video 2002: *Uncertainty*).

5) **Entangled particles and entangled minds**: Another issue of contention in quantum theory is whether entangled particles can communicate with each other over long distances instantaneously. This raises the issue of locality and non-locality of quantum effects and indicates that space may not exist in the micro world as we know it in the macro world. The argument has been framed in such a way as to say if there are only local effects, then there must be hidden variables. However, even if there are non-local quantum effects, then these variables are hidden as well because none of the four known physical forces can carry information faster than light. Of course, Einstein objected to the notion that particles could communicate over long distances faster than light. If there was faster-than-light communication, it would undermine the theory of special relativity. He insisted that there had to be hidden variables and that the particles that were supposedly entangled and show correlating characteristics at a distance must have had those properties before separating. Einstein and his graduate students designed the EPR thought experiment to reduce the idea of faster than light communication to absurdity. Later we will discuss how Bell proposed an experiment that would show that the instantaneous exchange of information between entangled particles could not be explained by local, hidden variables.

6) **Karl Popper, philosopher of science, challenged Bohr and Copenhagen interpretation:**
Karl Popper believed that Bohr's approach to quantum mechanics was not based in the philosophy of science and the scientific method. His main criticism of Bohr and the Copenhagen group was the introduction of *subjectivism* (the observer determines the outcome of the experiment) and *instrumentalism* (if the theory works, one does not need a rationale to show why it works). Popper believed in some form of EPR interpretation of quantum theory which Einstein espoused. He supported Einstein, whose theories, Popper believed, were falsifiable and had clear empirical support – thus they were scientific theories. Since Popper (1985) believed in scientific realism, (there is an objective reality that is independent of the human mind), he devised an experiment that would challenge quantum weirdness of entangled particles communicating faster than light at a distance (pp. 3-25). His experiment involved setting up a device that would entangle particles in the center, then shoot them in opposite directions through slits of varying widths to a photosensitive screen. In one trial, the slits would be the same width; in another trial,

the slits would be of different widths. Popper hypothesized that the diffraction pattern on the screen would be the same whether the particles were entangled or not. For example, if one slit was narrower than the other, the diffraction pattern from the narrower slit would be tighter than the diffraction pattern with the wider slit regardless of whether the particles were entangled. Popper said if the diffraction pattern for the particles going through the wide slit and the narrow slit are different, then the Copenhagen interpretation of faster than light action at a distance would be wrong since the particles did not communicate this difference to each other and behave the same way. On the other hand, if the diffraction pattern for the wide and narrow slits were the same, then the principle of entangled particles communicating faster than light at a distance would be true. Hence the experiment, as proposed, involved falsifiability of the hypothesis.

In 1983, the experiment was performed by Kim and Shih, and it confirmed Popper's prediction that entangled particles did not show the same diffraction pattern when one slit was wider than the other, thus indicating that the particles did not communicate with each other instantaneously as expected in the Heisenberg interpretation. However, Popper's experiment set off a firestorm in the physics community with defenders of the faith protecting their paradigm as Kuhn's philosophy of science predicts. Even the team (Kim and Shih) which conducted the experiment did not believe that Popper had scored a victory over the Copenhagen interpretation. They claimed that Heisenberg's uncertainty principle applies only to individual, unentangled particles, not to pairs of entangled particles (1999). What Kim and Shih are ignoring is that the whole idea of entanglement comes from the Heisenberg Uncertainty Principle which is certainly part of what is called the Copenhagen interpretation. Moreover, Bell's theorem, based on the assumptions of quantum physics, Copenhagen style, does indeed indicate that entangled twin particles affect each other at a distance. However, Bell's theorem also seems to contradict the Uncertainty Principle in that to pair and entangle particles takes the uncertainty out of their behavior such that their momentum and location, energy and time, spin and angle should thus be knowable simultaneously.

Is Kim and Shih's proviso (Copenhagen means single, not entangled particles) a legitimate fix or a fudge factor that they have inserted into the theory to protect it from falsification? The experts disagree, but I am sure that quantum physicists are likely to see Kim and Shih's caveat as a legitimate qualifier to show the domain within which the theory works – even though it contradicts Bell. To yours truly, the test was whether the particles would communicate with each other and adjust their paths to the screen to show the same pattern and match their partner particle as predicted from the entanglement hypothesis. Since their diffraction patterns did not match, the experiment shows that no such instantaneous communication took place at a distance. So, the success of the experiment seems to depend on what one thought the purpose of the experiment was – whether to test the uncertainly principle or the faster than light action at a distance principle. I think Popper meant to test the latter which would also refute the former.

Now another hedge to the "spooky action at a distance" hypothesis is that if the experimenter forces one of the entangled particles to change, it destroys the entanglement and causes decoherence. Thus, the hedgers for entanglement would say that Popper's experiment forced one of the pair into a different diffraction pattern, thus destroying the entanglement and nullifying the experiment. This argument would seem to make entanglement untestable since the only way to differentiate between local, hidden variables and action at a distance is to change particle A after entanglement to see if particle B changes to match particle A's behavior. After all, if Copenhagen is true and the particles are in an indefinite state until measured, then the change of particle A from indefinite to a definite state (say spin up) results in a change of particle B from indefinite to spin down. There is no way of getting around the fact that the measurement changes the state of both particles, so the argument that you can't force a change to a particle and maintain entanglement is refuted.

It appears from observing the interference patterns on a screen that the particle is typically at the crest of the wave where there is positive interference (resonance is a better word). This is where most of the dots are seen on the screen.

The question inevitably arises that if an observation is required to collapse a wave and make a particle real, then who observed the whole universe and caused the wave to collapse and put the universe in a definite state rather than existing and not-existing. John Wheeler has a fix for this. To learn of this other-worldly theory, read on.

Imaginary Debate between Dr. Quantum and Skeptic on Uncertainty Principle

Dr. Quantum: You can't know the position and velocity of a particle at the same time. There is an inverse relationship between the knowledge of one and knowledge of the other. The more you know about one, the less you know about the other. And, this dilemma is not because of the effect of the measuring instruments on the particle – the uncertainty principle is inherent in nature itself.

Skeptic: But to determine the position of the particle, you hit it with light and this collision with light changes the particle's velocity. Therefore, it is the limits of your instruments, not nature, which prevents you from knowing both variables simultaneously. The problem is that you cannot see into the subatomic world and the instruments you use to detect the particle interferes with it and changes it.

Dr. Quantum: Nonetheless, at the quantum level, you cannot separate the observer from the observed, both are part of the same system and therefore the observer is a part of nature also. So, this dilemma is inherent in the observer-observed unit of reality.

Skeptic: But science, at least classical science, is founded upon the idea of mind-independent, objective reality, and quantum mechanics has made all reality subjective. Quantum particles existed before humans evolved (or were created); therefore, the position and speed of particles occurred together before human observers were there to interfere with the simultaneity of these two variables (location and speed).

Dr. Quantum: Nevertheless, as John Wheeler has shown, human consciousness, going backward in time, has created the reality we observe. This reality did not exist before humans were here to observe it because it existed in a state of quantum uncertainty until humans actualized it by their observation.

Skeptic: Again, this preposterous idea that humans have god-like powers and created the universe with their consciousness violates the foundational principles of science, namely, rational-empiricism and mind-independent reality. The assumption of classical science has

QUANTUM QUANDARIES AND QUACKERY — Various Interpretations of QM

always been that cause precedes effect in time. Thus, such a theory or principle of "backwards in time cause and effect" is mysticism, not science. I doubt that there is any credible evidence to support Wheeler's outlandish idea, fanciful as it might be.

Quantum Contradiction
1>You can't know the position and momentum (speed) of a particle at the same time (uncertainty).
2>Therefore position and speed of a particle do not exist simultaneously.
3>But particles can be in two states at the same time (superposition).
4>Therefore a particle can have a position and speed at the same time (superposition).
5>From this logic, it would seem that superposition contradicts uncertainty.
6>The escape hatch for falsification is the mantra that quantum physics cannot be understood rationally. If that is true, then it is unfalsifiable and unscientific.

Confusion between Potential and Actual in Copenhagen
The Copenhagen Interpretation treats all possibilities as real in one instance and as potentials in another instance.
1>In this interpretation, a particle can really be here and there at the same time – not potentially here or there, but literally here and there.
2>However, a particle is here and there until an observation is made to actualize its position in one place at one time. So was the particle actually here and there at the same time or was it potentially here or there at the same time before the observation is made to force it to be in one place at one time.
3>In Schrodinger's thought experiment is the cat really dead and alive at the same time, or is it potentially dead or alive before an observation is made?

Extreme Epistemological Implications of the Copenhagen Interpretation
1>All matter is caused by human consciousness because consciousness creates the universe and even past reality according to the participatory anthropic principle of Wheeler.
2>If you can't measure or observe something, it doesn't exist. A particle that can't be detected doesn't exist until it manifests itself.
3>Nature at the quantum level is beyond logic – it is irrational or trans-rational.
4>Mystical unity: Everything is one, the perception of separateness is illusion. This is behind the dream for a Theory of Everything (TOE) – to find the mother of all particles.
5>The physical world is an illusion. When you get down to the elemental particle level, there is no there, there.

How Much Does Consciousness Affect Physical Reality in the Copenhagen Interpretation?

Does consciousness create physical reality or does it simply alter reality in some way?

The question arises as to how much and in what manner does consciousness affect physical reality. There are at least three ways in which consciousness is said to affect the material world.

1>Consciousness affects the state or condition of physical reality. For example, hitting an electron with gamma rays will change its location and velocity. In this scenario, there is no indication that affecting the state of the electron created the electron. Logic would say that it is not the consciousness of the experimenter that changed the location and velocity of the electron but that hitting it with gamma rays is what changed these properties. However, Bohr would contend that the observer cannot be separated from the observed in interacting in any way with material objects, so ultimately it was the observer's consciousness that affected the electron. Some might go so far as to say, that the electron didn't exist until it was observed indirectly using this method, so that would indicate that consciousness did create the electron.

2>Observing an object (even without directly tampering with it in a physical way) affects that object's state in some psychokinetic way. The effect is often stated as actualizing one reality from a number of possible realities (superposition). This would be considered "action at a distance" since the mind is operating on something without any known mediating physical force. An example would be observing a cat in a superpositional state, collapsing the wave, and making the cat either dead or alive. In this scenario, consciousness did not create the cat – the cat was already there – observation simply made it one way or the other. However, this effect of consciousness is different from pure psychokinesis in that the psychic causes something to be physically changed in the way s/he wants it changed. In quantum observation there is no indication that the observer has a wish for the experiment to turn out one way or the other.

3>Consciousness creates physical reality out of nothing. In Wheeler's delayed choice experiments, he concludes that the moon, indeed the universe, would not exist without humans to observe it. This creative potential of human consciousness has come to be known as the *participatory anthropic principle*. This effect is much more than just affecting the state of an object that already exists, it is creating a material thing out of nothing. This power is usually reserved for a god in many cosmologies. In Judaism and Christianity, Yahweh spoke the universe into existence; and in Hinduism, Brahma thinks the physical world into existence. Thus, the participatory anthropic principle endows man with god-like creative powers and even power over time in the ability to create things in past time.

The implication of all this business of consciousness affecting material reality, is that quantum theory trumps classical theory in determining reality. The reasoning goes that since everything is composed of quanta (packets of energy and atoms) that macro reality must behave in accordance with quantum laws. This concept, however, ignores the holism principle that the whole is sometimes greater than or different from the sum of the parts.

The Many-Worlds Interpretation – A Sober theory?

Gribben (2009) relates how Hugh Everett first conceived of the Many Worlds idea, probably under the influence of alcohol.

Following a party at which a considerable amount of sherry was consumed, Everett (and colleagues) amused themselves by dreaming up increasingly ridiculous implications of quantum puzzles, like Schrodinger's cat...Everett's big idea, initially tossed out as a joke, was "What if the wave function doesn't collapse? What if the superposition of states stays forever?" (pp. 24-25).

Everett didn't consider this notion to be a wild idea then, nor later when sober, and would subsequently propose that the wave doesn't collapse, and all probabilities of a quantum superposition are actualized by the splitting of the universe. Thus, the Many Worlds Interpretation attempts to solve the problem of the particle's indeterminate state by saying that all the particle's possible positions are accommodated by the universe splitting so that each possibility is realized in some parallel universe somewhere. So, if a quantum event (such as random numbers being chosen by particles being randomly emitted from a radioactive element) determines who wins the lottery, then the universe splits into millions of new universes so that everyone who entered the lottery wins in some universe, somewhere – since there was a statistical possibility that each entrant could win. John Wheeler, who initially supported this multiverse theory, ultimately withdrew his support because it "carries too much metaphysical baggage."

My critique would be that the multiverse theory violates the law of conservation of mass and energy and Einstein's $E=MC^2$ because if the universe were splitting a googolplex plus times every millisecond, wouldn't we see some diminution of the mass in the universe we live in? If the universe splits in half and Schrodinger's cat splits in half, the cat would have to replicate the lost half in each universe and, of course, the whole universe would have to replicate itself – if it were to remain equal to the original universe. If the universe doesn't replicate itself, then we would surely sense some lessening of the mass in the universe with so many splits, and we would continually be divided until we approach zero. Furthermore, wouldn't we begin to see some of these innumerable universes that are constantly splitting off? But wait, there is a hedge to rescue this problem too. Because of **decoherence** (loss of information by the scattering of particles), each of these new universes is inaccessible to us. So, this theory or interpretation, is unscientific because it can never be tested unless we can overcome decoherence and find at least one of these new universes somewhere in the cosmos. Thus, the theory is insulated from falsification. It is rather like some theories about aliens and why we don't have pictures of them taken by inductees armed with cell-phone cameras. The answer is much like decoherence – it is because the aliens can become invisible at will, and they can make their spaceships invisible as well – therefore no picture can be made of them. And so it is with the Many Worlds Theory, the other universes become invisible because of decoherence.

Maybe there is another theory that can rescue decoherence. The **holographic universe** theory says that all information of the universe is preserved on the surface of a black hole and the universe we experience is a 3-D holographic projection of this information held in 2-D. So maybe all the conditions of the experiment that produce so many universes are preserved on the surface of a black hole. As stated, good scientific theories are set up in such a way as to be falsifiable. This one doesn't meet the test despite all the mathematical acrobatics that would make it so. However, there is one test which the many-worlders have overlooked. More on that later.

Furthermore, the many worlds interpretaton is a quantum leap of cosmic proportions. How is it that tiny, submicroscopic particles in superposition (many probable states) split off to create whole new universes, the size of the one we live in, which is so large that we cannot begin to measure its dimensions?

Moreover, how is it that each interpretation of quantum mechanics is said to be equally valid because they produce more or less equally accurate predictions even though they obviously contradict each other. Gribbin (2009) says:

If there was anything better than the Copenhagen Interpretation, it would have been discarded long ago. But there isn't anything better. There are only alternative interpretations, which are precisely as good as the Copenhagen Interpretation, in the sense that they are just as good at predicting the outcomes of quantum experiments (p. 23).

That same John Gribbin (2009) says that the Copenhagen Interpretation is not as good as the Many Worlds Theory in describing the multiverse.

But as Susskind emphasizes, the collapse of the wave function is not part of the mathematics of quantum physics, it is 'something that Bohr had to tack on in order to end the experiment with an observation'. Since the universe does not end with an observation, the Copenhagen Interpretation, quite apart from its other flaws, is inappropriate when we are trying to describe the Multiverse (p. 171).

Saying that each alternative interpretation of quantum mechanics is equally valid because they are just as good at predicting the outcomes of quantum experiments is like saying that the Ptolemaic geocentric view of the universe is equally as valid as Copernicus's heliocentric theory because they both predict celestial motion even though they contradict each other. In the same way, how can the Copenhagen interpretation and Many World's interpretation be equally valid when they contradict each other. One might reasonably ask if these theoretical interpretations contradict each other, how could their mathematics render the same predictions? It would seem to me that for contradictory theories to converge on the same answers mathematically, there would have to be fudge factors built in that would force the mathematics to convergence with such divergent concepts and assumptions. As indicated previously, shooting particles through double slits and getting a pattern of dots on the screen which suggest wave interference would enable physicists to predict the patterns next time the experiment was tried. However, their interpretation as to what goes on between shooting the particles and their patterns on the screen are widely divergent and even contradictory. Be that as it may, this claim is a typical thought process in the culture of physicists, i.e., to make contradictory claims and then claim there is no contradiction, thus having their proverbial cake and eating it too. The Copenhagen interpretation and the Many World's Interpretation cannot possibly be reconciled conceptually even though they might make equally accurate mathematical predictions by inserting constants or patches. Here are their irreconcilable differences.

Copenhagen Interpretation vs. Many Worlds Interpretation
Probability wave collapses vs. Probability wave doesn't collapse
Cat becomes dead-or-alive in one universe vs. the cat is dead-and-alive, but in multiple universes
Observation causes cat to be dead or alive vs. consciousness plays no role in splitting universe
Particles have no definite state until observed vs. particles have a definite objective state
Faster than light communication vs. nothing faster than light

The argument that these alternative interpretations do not contradict each other is tantamount to saying that a cat is dead and alive simultaneously and that there is no contradiction in such a claim. As stated before, if the math for each interpretation works out equally well, then it must mean that the math has been finagled to fit each interpretation; because the various interpretations obviously contradict each other. A wave cannot collapse and not collapse at the same time unless one lives in Eastern philosophy where contradictions can coexist.

Now, not only does Everett create a fix for why we mortals cannot see the universe splitting, but he also creates a fix for why we don't feel all this constant splitting – especially as our bodies are splitting countless times every nanosecond or so. He says that this "feeling" question is like the criticism of the Copernican theory of heliocentrism in which the earth spins and orbits the sun. Since we don't feel the spinning of the earth and its motion around the sun, common sense would say that the earth is stationary. However, this counterpoint by Everett is easily put to bed now that humans have visited the moon. Since the moon is in a locked orbit so that its spin is synchronized with its orbit around the earth, it spins once in 27 days as it completes one orbit in 27 days, and the moon's orbit is in the same direction of the earth's spin. This means that the lighted face of the moon is always pointing toward the earth. Hence with time lapse photographs taken from the light side of the moon, it could be seen that the border between dark and light (penumbra) would return to its original observed position on the earth in 24 hours, thus indicating that the earth spins every 24 hours. This spin of the earth would indicate that the apparent rotation of the heavens around the earth is incorrect because it is the earth's spin that makes the heavens appear to orbit around it. No such observation could be made regarding MWI which would show that the universe splits and forms many universes countless times every fraction of a second.

Moreover, several lines of theory are pointing to a multiverse including eternal inflation, string theory, dark energy, and of course the Many Worlds theory described here. For this, and other reasons, there is a defection of some physicists from the Copenhagen Interpretation to the Many Worlds theory. Another reason is that Copenhagen mysticism has become an embarrassment to some physicists who want to keep physics looking like real science instead of Eastern mysticism. Here is what Gribbin says about this trend:

But as Susskind emphasizes, the collapse of the wave function is not part of the mathematics of quantum physics, it is 'something that Bohr had to tack on in order to end the experiment with an observation'. Since the universe does not end with an observation, the Copenhagen Interpretation, quite apart from its other flaws, is inappropriate when we are trying to describe the Multiverse (p.171).

> The reason that consciousness is said not to be necessary in the Many Worlds Interpretation is that the wave does not collapse as in the Copenhagen view. However, the wave is said to decohere after the universe splits so that the universes cannot communicate with each other. By indicating that consciousness does not determine physical reality, the Many Worlds Theory does rescues quantum mechanics from mysticism, but at what cost? **Which is easier to believe – that consciousness created the universe or that superpositions (multiple probabilities of an event) create innumerable universes every microsecond?**
> ~Another critique of the Many Worlds Theory is that if quantum superpositions create different universes for each possibility, then there would be no interference patterns on the screen since the superposition would have dissolved into one state in this universe, and since the universes cannot communicate with each other because of decoherence, there would be no interference between one universe and the other since they are completely separate.

The Quantum, 2-Bit Computer and the Many Worlds Interpretation

As demonstrated earlier, theory and technology are not always correlated on a one-to-one basis. As we have seen with much of electronic-photonic technology, inventions were made before Maxwell's mathematical description of electro-magnetism in the 1800s and the development of quantum theory in the 1920s. If quantum computers are developed and become a part of our technological landscape, which interpretation of quantum mechanics will the computer operate on? Deutsch (1998) and Gribbin (2009) have argued that a quantum computer will operate on the Many Worlds interpretation and that such a computer is strong evidence (if not proof) of the multiverse. Since the Many Worlds interpretation contradicts the Copenhagen interpretation (wave collapses vs. wave doesn't collapse, one universe vs. many universes), then both interpretations cannot describe the principles undergirding quantum computing.

Actually, in a sense, all computers are quantum computers since they operate on quantum particles, i.e., electrons and photons. A conventional computer operates on the principle that an electronic or light signal can be blocked by a switch (thus creating a zero) or allowed to pass (thus creating a 1). However, a quantum computer would theoretically increase computer speed and the amount of data that can be handled at once because of *superposition*. As stated a conventional computer operates on base 2, that is, information comes in two basic bits, zero or one (i.e., present or absent). Hence the number of bits increases as a doubling geometric progression: 2, 4, 8, 16, 32, etc. Similarly, a quantum computer would operate not only on binary bits or qubits of information (0 and 1), but in superposition which means the wave can carry both 0 **and** 1, not just 0 **or** 1 as in conventional computers. In other words, the qubit involves two bits of information carried at the same time, rather than 1 in the conventional computer. The following is a summary-analysis of the arguments offered by Deutsch (1998) and Gribbin (2009) for the connection between the Many Worlds interpretation and quantum computing.

1> Because of superposition, a quantum computer can double the bits or qubits that can be carried as in a double-slit type, binary superposition. It can theoretically carry much more data if the number of superposition states is greater than two as with a slit experiment with more than two slits.

2> The Copenhagen interpretation would not work in a quantum computer because once you measure the wave to extract the data, the wave collapses to one state and half the data would be lost - then you are back to a regular computer. To wit, if a wave is carrying both 0 and 1, the collapse would make it either zero or one, not both; therefore only half the data is left.

3> If there is a zero there is no information; therefore, no wave at all since a quantum wave is a field of information. In Quantum Mechanics, it is alleged that an entity can be in more than one state at a time, but not in a state of existence and non-existence at the same time. For example, a particle can be here and there simultaneously and a cat can be dead and alive, but a wave cannot tell you that a particle exists and doesn't exist or a cat exists and doesn't exist – after all, a dead cat still has a physical existence.

4> The Many Worlds, as originally conceived, would not work because once the universe splits, the 0 goes into one universe and the 1 goes into the other, and, because of decoherence, they cannot be brought back together or communicate with each other. Again, half the information is lost to each world.

5> The only way a quantum computer would work *in theory* is with *Deutsch's modified Many Worlds Interpretation* in which universes can come back together after the split. Deutsch argues that universes can come back together after a split and produce positive interference so that the two halves of the data come back together. However, the collective data of the two universes is in a superpositional state so that if the wave is collapsed (or decohered), again half the data would be lost. The only way to rescue the data then is with a conventional computer to calculate the data once half of it is rescued. By having longer strings of binary data in the quantum computer, this would give some advantage over 1-bit, conventional computers.

6>Another problem is that a quantum computer assumes that a particle is in superposition (all possible states at once carrying info in each state). While the Copenhagen interpretation and the Many Worlds makes this superposition assumption, Bohmian Mechanics and QBism does not. Hence a quantum computer would not work if Bohm and QBism are right and there is no superposition. Thus, if a quantum computer is not produced, it would indicate that Bohmian mechanics and others which do not assume superposition would be supported. In Bohmian mechanics, the particle is always in a definite position, but the wave is said to be in a superposition containing all the data about possible positions and momenta. However, not all of this data are active at once, each datum is potentially active. It is like a library that has myriad information, but I (as a single person, like a single particle) cannot activate all that information at once. I can activate it sequentially but not simultaneously. Similarly, in the lottery, all numbers can possibly be played, but only certain ones are played, and only one combination can win the big prize. Possibility is like potential energy - it is stored but not active. Kinetic energy is energy that is activated.

In information science, data units are based on distinction and disambiguation. Therefore, to be meaningful, a bit of data must be either 0 or 1, not both 0 and 1. In linguistics, we learn that a phoneme is a <u>bit</u> of information in language. The phoneme is a difference in sound that makes a difference in meaning. For example, if the phoneme "b" is substituted for the phoneme "p", the meaning of a word will change in English and many other languages. To wit, <u>b</u>it has a very different meaning from <u>p</u>it because of the phonemic distinction between p and b. If the English speaker says pbit (in which the p and b are superimposed on each other) there is no meaning. Similarly, if 0 and 1 are interchanged in a number, the value of that number is changed. If the uncertainly principle is true, then when a measurement is made, one cannot know whether s/he will get a 0 or 1 and therefore cannot rely on the accuracy of a calculation. Perhaps, there are theoretical hedges to get around the problems I have identified, but the quantum computer has been worked on since 1998 with little success (Gribbin 2009: p.73). What we will probably see when a quantum theory is projected into the macro world in the form of a computer, is that we get the same absurd answer as a zombie cat (a cat that is both dead and alive) when the Copenhagen interpretation is brought into the classical realm. Gribbin (2009) informs us that in 2001, a team at IBM's Almaden Research Center actually factored the number 15 (into 3x5) using Shor's algorithm. This soggy quantum computer is described as

...a molecule which contains five fluorine atoms and two carbon atoms, giving them seven nuclear spins to play with – seven qubits. But not just one molecule – they used a solution containing an estimated billion-billion of these molecules in about a thimble full of liquid monitored by Nuclear Magnetic Resonance with the averaging of all those molecules effectively compensating for any errors introduced by decoherence. This proved that quantum computing works, proved that Shor's algorithm works, and makes it difficult to doubt the existence of the

multiverse (p. 73).

The problem with this analysis is that Gribbin does not offer us any specific explanation of how this infant quantum computer reflects superposition of seven states of the fluorine-carbon solution, how this shows the universe splits into each state, how the seven universes come back together to interfere with each other, and how the answer of 3x5 = 15 is extracted from the resulting interference pattern without collapsing or decohering the wave.

Now, there seems to be an international race, similar to the space race of the 1960s, to develop a workable quantum computer with more and more qubits. The latest development in quantum computing was revealed in the summer of 2017 at the International Conference on Quantum Technologies, held in Moscow, Russia. The co-founder of the Russian Quantum Center and head of the Lukin Group of the Quantum Optics Laboratory at Harvard University, Mikhail Lukin, announced that his team had successfully built a 51-qubit quantum computer. Professor Lukin said that this new development involved the use of time crystals (rather than a soggy, liquid computer) (Caughill 2017):

Basically, the unique thing that happens with these time crystals is that they can be a stable state of matter. These states, in principle, can hold quantum coherence for a long time. So basically, it means you can have super-positions of states. That's kind of the basic ingredient for all this quantum science and technology. So on one hand, we can think about using it as memory for a quantum computer—in principle it's true, but as for practical use...it's not so clear.

Another development in quantum computing is that entanglement of particles can be used to convey information despite the fact, for years, theoretical physicists have told us that entanglement cannot be harnessed to develop faster than light communication.

Nutshell: I suspect that it might be possible for photons and electrons to carry information in a computer since these quantum particles can carry radio and television signals. However, I don't see how such technology would involve the preposterous idea of the universe splitting and converging. I think the technology could work without this superfluous, science-fiction like idea. Moreover, making the Many Worlds work with quantum computing requires a major overhaul of the original theory, to wit, it requires a nullification of Everett's idea that humpty-dumpty split universes can never be put together again. Instead, Deutsch's hedge requires that these split universes reconvene and interfere with each other. I think Gribbin and Deutsch are a long way from proving their theory of a universe with a split personality – or perhaps multiple personalities - using quantum computing as evidence. Because of the tenuous nature of quantum phenomena, quantum computing may go the way of cold fusion, perpetual-motion machines, Super Symmetry, String Theory and contracting-expanding space to achieve near luminal speeds.

QUANTUM QUANDARIES AND QUACKERY   Various Interpretations of QM

Furthermore, superposition is sometimes phrased that the particle is "neither here nor there", rather than "both here and there". If the particle is neither here nor there, it is not really in superposition and therefore could not carry any information as opposed to carrying double the information.

The Challenge of Quantum Computing
1> Placing particles in superpositional states to multiply the amount of information carried.
2> Extracting each unit of information from the superposition without collapsing or decohering the wave so that half (or perhaps all) the information is lost.

Bell's theorem of inequalities as related to quantum entanglement

Nutshell: Bell's inequalities is like an ESP test in which the subject must beat chance in guessing the right card. Since there are 5 types of Zener cards in the Rhine ESP test, guessing the right card by chance would be 20% (1 in 5). If a subject scores significantly above 20%, there is support for ESP.

What is Entanglement? Entanglement occurs when two particles are created[1] together in such a way that they have linked properties that are exactly opposite or mirror images of each other – thus they have reverse symmetry. EPR (Einstein, et al) assume that the particles have these matching properties when formed locally; but Quantum Mechanics (Copenhagen Interpretation) assumes that the particles have no definite state or perhaps have every possible state (superposition) until they are measured at some distance and time after entanglement. The problem with QM is that the particles must have some complementary DNA planted at birth for them to show opposite traits (such as opposite, spin, opposite axis, matter/antimatter) when measured later – otherwise their behavior would not be correlated and entanglement would have no meaning.

[1] Created is not an accurate description because particles created from nothing would violate the law of conservation of mass and energy. More accurately, the particles are said to condense from the energy field that pervades space so that energy transforms to matter to form point particles. Thus, a more accurate picture is that energy is converted to matter particles.
Why must entangled particles conserve angular momentum and other conservation laws?
Because, again, matter and energy can neither be created nor destroyed, so that all conversions from mass to energy and energy to mass must net zero changes in the mass-energy sum. So, when twin particles are formed, one matter, the other anti-matter, their algebraic sums must equal zero:
matter + anti-matter = zero
clockwise spin + counterclockwise spin = 0
axis of spin - opposite axis of spin = 0 (for example +90-degee axis + -90-degree axis)

Bell's Inequalities were a mystery to me. No writer that I had read would say what the *inequality* in Bell's theorem is until I encountered Bob Eagle (2012). What is unequal? According to Eagle, the inequality refers to the unequal results an experiment might produce if one starts with Einstein's assumptions - that particles acquire matching (inversely-correlated) properties when locally entangled. Essentially, the theorem indicates that the EPR (Einstein, *et al*) prediction is that 1/3 of entangled particles are expected to show matching behavior based on chance alone. If 33% is chance expectation and if there is a statistically-significant deviation above 33%, then the matching behavior is due to local causes or pre-existing information in the particles. If the matching behavior is significantly less than 33% chance, then the matching result must be accounted for by non-local factors that communicate instantaneous (faster than light). Therefore, Bell's inequalities or deviation from chance would indicate that if hidden variables are allowed in order to create the match between separated, entangled particles, then the variables have to be non-local if the experiment is set up in such a way that the particles could not have obtained the information to match behavior while in close proximity. *Bell's thought experiment involves changing the behavior of particles after being separated by some distance to see if the particles match or give complementary responses instantly.*

As noted, Bell's theorem is a refutation of the EPR challenge and hypothesizes that entangled particles communicate non-locally faster than light. Unless particles are entangled, they show unrelatedness in their spin axes, polarization, etc. and are not correlated as they are if entangled. Einstein *et al* had said that apparent correlation of entangled particles indicated that they had those attributes to begin with and that correspondence was not due to faster than light communication between them. Although Bell did not have the technology for testing the matching behavior of entangled particles at a distance, Alan Aspect and others later tested this theory with precision technology.

Kaiser (2014) explains Bell's theorem in this way:

Bell's theorem indicated that particles show independence in their spin unless entangled and that entanglement cannot be explained by local hidden variables, but might be explained by non-local Bohmian variables. In his article, Bell demonstrated that the mathematics of quantum theory require entanglement; the strange connectedness is an inescapable feature of the equations. But Bell's proof was purely mathematical and didn't show that nature actually behaved that way, only that physicists' equations did. The question remained: Does quantum entanglement occur in the real world? (SR10).

Procedure for Testing Bell's Theorem and Entanglement:
Copenhagen, EPR and MWI Perspectives

1~Two quantum detector machines are set up some distance apart (theoretically distance does not matter for communication between entangled particles).
2~The machines are set up to detect some quantum property of entangled particles such as spin

axis or polarization. The assumption here is that spinning, charged particles, such as electrons or protons, create a tiny magnet with a South Pole and North Pole. If the particle is spin up, it would mean that the North Pole is spinning in the opposite direction of the spin-down particle. The spin-down particle would have its North Pole 180-degrees from the spin-up particle. Of course, the photon does not have charge, so photons would be described as polarized up or down. However, some contend that entangled photons are polarized at 90-degree angles from each other. This 90-degree angle is necessary since electromagnetic waves vibrate back and forth with the forward vibration 180 degrees from the backward vibration. Therefore, a wave that is vibrating at 0-degrees will also vibrate at 180-degrees in the opposite direction. Hence a photon that is polarized at 0 is also polarized at 180, so entanglement has to involve photons polarized at 90 degrees apart, not 180 degrees.

3~In the Aspect experiment using photons, the detector is set to detect three polarization axes or angles, i.e., 0, 120 and 240.

4~If one of the twin photons is found to be polarized vertically, the other particle will be measured to be polarized horizontally (90-degrees from the twin particle). On the other hand, in spin experiments with fermions (such as electrons), if one particle is found to spin clockwise, the other will be spinning counterclockwise thus preserving angular momentum in which the spin of one particle exactly cancels the spin of the other particle and the net spin is zero.

5~In the Copenhagen interpretation, the machine does not completely determine the orientation of the particle - the detector causes the wave, which is in an uncertain state of spin up or spin down, to collapse and actualizes one of the two possible states the particle could be in. When the wave of particle A collapses, twin particle B is then compelled to collapse in the opposite, complementary state so that conservation laws can be obeyed and the resultant combination of the two states can be net zero. Older interpretations would indicate that the consciousness of the experimenter causes the collapse, but more recent interpretations might say that it is the measurement by the machine (which is assumed not to have consciousness) that collapses the wave.

6~In the EPR interpretation, Einstein *et al* would say that the particles acquired their matching, complementary orientations when they were entangled, and the machine simply confirms this local determinism when the measurement is made. However, since this process of matching complementary properties cannot be observed, they are called *hidden* variables. Nevertheless, an EPR advocate might argue that when twin complementary particles (one matter and the other anti-matter) are hatched in the quantum foam, they have inherently complementary properties. This complementarity can be seen when they attract because of opposite charge and annihilate each other (actually they convert each other to pure energy in the form of photonic radiation) when they come together. Thus, one might say the wave collapse occurs when the annihilation occurs. In any case the net result is zero matter and the law of conservation is thus preserved. If such twin particles have matter/anti-matter complementarity, then they could have other inherent complementarities such as opposite spin, axis orientation, etc. Feynman says that matter/anti-matter particles also have opposite time orientations since the matter particle goes forward in time and the anti-matter particle travels backward in time (Clegg 2014: p. 70).

Despite this admission that twin particles entangled in the quantum foam have inherent complementary properties, the tests of Bell's Theorem supposedly prove that the correlating behavior of entangled particles at a distance is not due to pre-existing, innate information that was locally acquired in the entanglement process. Thus, these entanglement experiments have supposedly eliminated local, hidden variables as a determinant of the correlation of entangled particles. The argument to support non-local determination is that the first particle is changed after the entanglement when it is measured, and the other particle follows suit instantly. This seems to me to be rather like Pauli's Exclusion principle which dictates that electrons with the same properties (say two spin-up electrons) cannot inhabit the same orbital. Instead the electrons must have complementary properties. If one is spin-up, the other must be spin-down to live in the same part of the shell. This fact implies that electrons already have this information in their DNA.

*In Popper's experiment with entangled particles, he varied the width of the slit (in a double slit experiment) for the two particles. Thus, he tried to determine the wave behavior of the particle. Contrary to entanglement prediction, Popper expected that the twin particle would not manifest the same behavior if it went through a slit of a different width. What Popper found was that the entangled particles did not demonstrate the same wave diffraction or spread if the slits were of different widths, contrary to the expectation from other entanglement experiments, which indicate that the twin particle will always match the behavior of the other.

According to Kaiser's article, the experimenters will attempt to close every loophole (or local hidden variable) that could cause the apparent entanglement. One loophole is that somehow the machine itself may send messages to both particles to cause them to correlate their behavior. Since both detectors have to be set by the human experimenter, the machine could somehow exert an effect on the human setting the detectors. Have computers, like Hal in *2001: A Space Odyssey*, really become conscious and intentional.

To close this intentional-machine loophole, the experimenters have decided to use ancient starlight from distant quasars that are too distant from each other to have been entangled. If light arrives at an odd microsecond, one setting will be used, but if light arrives at an even microsecond, then another setting will be used. This should give a random flip of the coin that would obviate any machine influence.

If, as we expect, the usual predictions from quantum theory are borne out in this experiment, we will have constrained various alternative theories as much as physically possible in our universe. If not, that would point toward a profoundly new physics. (Kaiser 2014: SR10).

Alain Aspect (Jorlunde Film 1985) had another method to ensure that the machine or the operator did not determine the complementary behavior of the particles. He installed an optical switch that could, with a rotating mirror, change the orientation of the filter faster than a signal traveling at the speed of light could reach both particles at opposite ends of a 12-meter tube - where the particles would be when they mutually decided which polarization to take. Of course, the experiment showed that quantum speed is instantaneous so that a quantum signal from some other source, traveling faster than light, could conceivably affect the polarization of the two particles. However, this argument could be countered by saying that only entangled particles can communicate at quantum speed (infinite speed) at any distance. Initial entanglement surely imparts some information channel between twin particles that randomly-selected, independent particles do not have. The minimal information that it could create in the particles is a channel of communication between them.

The following is a graphic demonstrating how Bell's Theorem was tested by Alain Aspect and others.

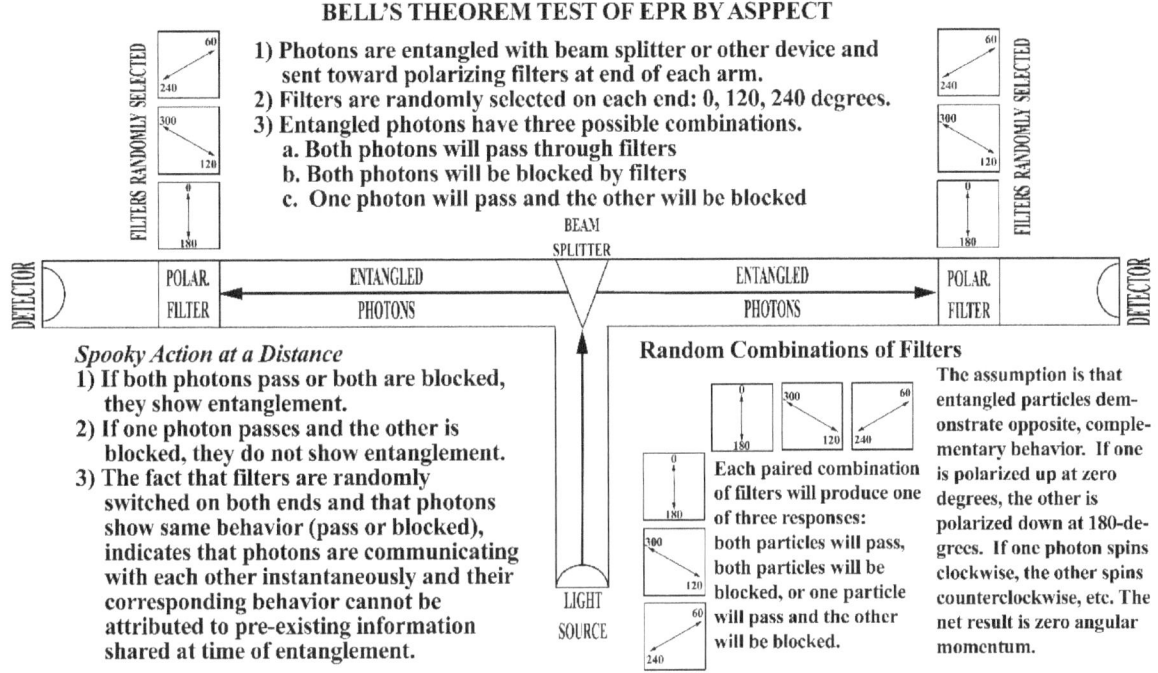

*One correction to the table above is that in the case of photons, entangled particles are said to be entangled at 90-degrees from each other since photons vibrate in opposite direction. Thus, a photon polarized at 0-degrees is also polarized at 180-degrees since it vibrates in both directions. On the other hand, when dealing with fermions such as electrons, opposing spin is said to be 180-degrees apart so that in an entangled pair of electrons, if one is spin up, the twin electron is spin down (180-degrees apart). So, boson pairs (such as photons) are 90-degrees apart and fermion pairs are 180-degrees apart.

Nutshell: The key to understanding the experiments to test Bell's Theorem is proof by contradiction: If entangled particles show the same or complementary behavior to new information at a distance from each other, it cannot be due to local, hidden variables. Therefore, it must be due to instant communication at a distance between particles at the time of measurement. There are two ways to test for non-localism: 1) randomly match three types of polarizing filters that each particle must either pass through or be blocked. If the particles have pre-existing information, then they will match each other's behavior greater than chance expectation, if no pre-existing information, they will not beat chance; 2) Switch filters after the twin particles have been emitted, so that the particles could not have known beforehand what combination of filters there would be.

The following table shows the logic of Bell's inequalities and how the probabilities of EPR must be equal to or greater than chance expectation (based on Eagle 2012). There are three possibilities this experiment might be designed to check: 1) What result would occur if there is no embedded information in entangled particles, 2) What would be the result if there is information embedded in entangled particles, but no information is exchanged after entanglement, and 3) What would be the result if entangled particles with embedded information can communicate instantaneously and change the information after embedded and separated? Since there are three polarizing filters, if there is no information embedded in the particles, the chances are 1 in 3 of getting a matching response with any two combinations of filters. If there is locally embedded information, then the entangled particles should exceed 33% random probability. If there is locally embedded information and continuous, instantaneous communication between particles, then the particles should show matching behavior which far exceeds chance even when the filter orientations are changed after the particles are entangled and separated. The following graphic is the way Bell approached this thought experiment according to Dr. Physics (Eagle 2012).

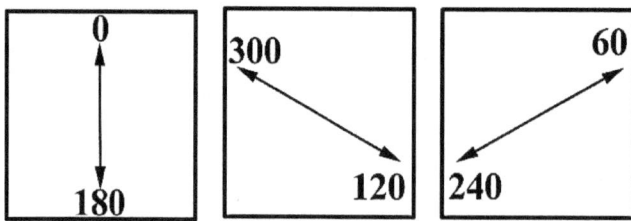

Probabilities of Pass (P) or No Pass (X) thru filter   Random Filter Combinations

(1) Vertical Polarizer	(2) 120-degree Polarizer	(3) 240-degree Polarizer	Filters 1 & 2	Filters 2 & 3	Filters 1 & 3
P	P	P	~~same~~	~~same~~	~~same~~
P	P	X	same	different	different
P	X	P	different	different	same
P	X	X	different	same	different
X	P	P	different	same	different
X	P	X	different	different	same
X	X	P	same	different	different
X	X	X	~~same~~	~~same~~	~~same~~

The following points summarize the logic of this table, based on Bell's Inequalities:
1~ Eliminate the top and bottom rows of the above table on right (in white background) in which there is the same particle behavior (both pass or both blocked) in all three combinations of filters. Same behavior would indicate that particles have same information already. As to why the bottom and top rows are eliminated is a mystery to me since that would also be experimental information about the matching behavior of the particles.

2~ The other six rows of the white table show that EPR would predict that the twin particles should show the same response (both particles pass or not-pass) 1/3 (33%) of the time based on the probability that the particles contain information on all these properties. Since 33% is the expected matching percentage if EPR is true, the experimental result would have to be equal to or greater than 33% to support EPR. Bell's inequalities indicate that the experimental result must be unequal to chance expectation – a rather backward way of stating the case. However, experiment shows that the *same responses* by both particles (both pass, both blocked) occurs only 25% of the time. So, if Bell's inequality requires that the same behavior must be 33% or greater, the 25% falls 8+% short of the EPR chance expectation and violates Bell's inequalities. However, I would question why rows 1 and 8 (where the particles show the same behavior in all three combinations) would be eliminated - except to skew the results. If rows 1 and 8 are added back in, then the table shows that 50% of the time, the twin particles show the same behavior. Since Bell's inequalities require that 33% or more of the time the particles match, then 50% matches do not violate Bell's inequalities. I would also question why the 8% deviation from chance, which could be a significant deviation from chance, should be disregarded. Instead of exceeding chance expectation, the experiment shows the matching is significantly below chance expectation. Why would the particles mismatch themselves below chance level? Either way there is a violation of chance and randomness. It doesn't matter whether a coin flip comes out 58% heads or 58% tails, there is a deviation from chance, and in enough trials an 8% deviation could be statistically significant.

There is a much simpler, more direct way of presenting this concept of testing local determinism (EPR).

Sometimes, in science, one method used in testing an hypothesis will yield a different result from another method that is used. I am proposing the following methods of testing entanglement and instant communication between particles.

Simplified Entanglement Experiment

The following is a simplified version of the Bell Theorem and a proposed experimental design for testing it. The experiment is based on three assumptions as to where information in and between particles resides:
1~**Intrinsic information in single particles**: There is information in the DNA of individual, unentangled particles, such as, how photons or beams can be polarized on various axes. It is known from experiments in which light is polarized on one axis will be blocked by a filter oriented at 90 degrees to the original polarization, but a filter at a 45-degree angle will allow 50% of the light to pass. However, some of the light that is blocked by a 90-degree filter can be recovered by placing a 45-degree filter between the two filters oriented at 90 degrees to each other.
2~**Acquired information when particles are entangled**. There must be some information

embedded into entangled pairs of particles locally for them to behave as a unit. For example, there has to be a channel established between them if they are to communicate instantaneously at a distance at some later time. What would entanglement mean if it does not involve the sharing of some information at the time of entanglement?

3~**Acquired information at a distance after particles are entangled:** The Copenhagen argument is that when one particle signals another regarding its quantum state (such as spin up) at a distance, the other particle assumes an opposite, complementary state (such as spin down).

Statistical Expectations Based on Known Laws of Polarization

Assumptions:
1>Copenhagen assumption is that entangled photons do not have definite polarization until measured. This contradicts Bohmian mechanics which indicates that photons do have definite polarization before measured, but they can communicate also non-locally according to Bohm. The hypothesis to be tested here is that the photon does have a definite polarization before measured and that the filter does not just passively measure the polarization of a photon, the filter actually can change the polarization of the photon. To wit, if unpolarized light is passed through a vertical filter, 50% of the photons will be absorbed and 50% will pass through as *vertically polarized* light. If the filter only measured photons that were vertically polarized, then perhaps less that 1% would pass because only a small percentage of unpolarized light would be vibrating in an exactly vertical direction. Similarly, when vertically polarized light passes through a 45-degree filter, 50% of the vertically polarized light will become polarized at a 45 angle and 50% will be absorbed. So, the filter is not just passively measuring photons, it is an active instrument that can change polarization. This means that the individual photon in unpolarized light already had a pre-existing polarization and was vibrating at a definite angle before measured. However, a beam of unpolarized light had photons vibrating at every conceivable angle. Thus, Copenhagen confuses an aggregate beam of unpolarized light which has every conceivable polarization with an individual photon which has only one polarization, not every possible polarization before measured. Once measured by a filter, the photon can change its angle of polarization.

2>If Photon A is polarized in any direction, twin photon B will instantly become polarized at a 90-degree angle from Photo A. Therefore, if each photon encounters a vertical filter, the vertically-polarized photon will pass, but the horizontally-polarized photon will be blocked.

3> Symbols used: [|] = vertical (0/180) filter   [-] = horizontal (90/270) filter   [/] = diagonal or 45 filter

4>If Photon A does not pass through the filters, then the statistical expectations would be the same as in the table because Photon B would be in the opposite position and would pass or not pass based on the same statistics; therefore the assumption in the table is that photon A passes through the filters indicated.

5>The prediction is that if tests are run and using the following combination of filters, the photon pairs should follow these statistics very closely based on the known laws of polarization. The degree to which the actual observations of passage/blockage deviate from these expected percentages would indicate whether there is true entanglement or not. If there was a radical departure from these expectations, then entanglement would not be supported.

Table of Predictions of Passage Based on Known Probabilities of Polarization			
If Photon A Passes	Photon B Passage Rate if Photon A Passes		
	[|]	[-]	[/]
[|]	[||] 0%	[|-] 100%	[|/] 50%
[-]	[-|] 100%	[--] 0%	[-/] 50%
/	[/|] 50%	[/-] 50%	[//] 0%

Predicted Outcomes from Table Above assuming that particle pairs are 90-degrees apart.
1> [\|\|] combo: If A is polarized vertically, A would pass, but then B would be polarized horizontally and would be blocked by vertical filter.
2> [\|-] combo: If A is polarized vertically, it would pass; then B will pass through a horizontal filter 100% of time.
3> [\|/] combo: If A is polarized vertically, it will pass; then B will have a 50% chance of passing through a 45 diagonal.
4> [-\|] combo: If A is polarized horizontally, it will pass, and B will pass because it is polarized vertically.
5> [- -] combo: If A is polarized horizontally, it will pass, but B will not pass horizontal filter because it is vertically-polarized – 0%
6> [-/] combo: If A is polarized horizontally, B will pass a 45 filter 50% of time
7> [/\|] combo: If A is polarized at 45, it will pass, and B will pass vertical filter 50% of time
8> [/-] combo: If A is polarized at 45, it will pass, and B will pass horizontal filter 50% of time
9> [//] combo: If A is polarized at 45, it will pass and B will not pass a 45 degree filter because it is polarized 90 degrees from A or 135 degrees.

If all photons that pass were run through a second filter of 45 degrees from their polarization, the odds would be 50% that the photon would pass the 45.

After many trials, the actual percentages could be placed in the table and the difference between the expected percentage and actual percentage could be calculated. A chi-square test would indicate the level of significance or probability of entanglement (.01. .05, etc.).

My prediction is that the results would turn out like the Popper experiment in which entangled photons sent through different size slits did not show the same interference patterns thus supporting EPR rather than Bell inequalities. I predict that entangled particles would not conform to non-local, "spooky-action-at-a-distance" but would support local, hidden variables. It is possible that particles would show independence and not demonstrate entanglement at all.

Strong Entanglement: Beam of Entangled Electrons deflected by Magnet: I would like to propose an experiment that is similar to the J.J. Thompson experiment involving a stream of electrons. In this experiment a stream of electrons was generated in a vacuum tube. When a horse shoe magnet was placed over the tube, the beam was deflected in one direction and when the magnet's north and south poles were reversed, the beam was deflected in the opposite direction. Similarly, in the Stern-Gerlach experiment, electrons were either deflected up or down (or at some other angle) when run between magnets thus indicating the hypothetical spin direction of electrons which actually means magnetic polarization. Since an electron has a negative charge only, it is difficult to see why it would have any polarity at all. Perhaps magnetic polarization has nothing to do with positive and negative charges – although we generally think of polarization as involving opposing charges. As to why J.J. Thompson's experiment produce only one direction of deflection and the Stern-Gerlach experiment produced multiple directions of deflection is a mystery to yours truly.

My proposal is to create entangled beams of electrons going in opposite directions in a tube. Then place a magnet at one end to deflect that beam in a certain direction. If there is strong entanglement, the beam going in the other direction should be deflected in the opposite direction from its twin. Then reverse the magnet to see if the opposite beam changes the direction of deflection. However, in delayed choice quantum eraser experiments, the deflection of one twin of the pair by a prism or mirror, did not change the direction of the other twin. Likewise, in the Popper experiment, the change in the diffraction of one twin by going through a different size slit did not affect the diffraction pattern of the other twin, so it is doubtful that entanglement is a strong enough factor for one twin to change the trajectory of its counterpart by deflection with a magnet. Apparently, quantum entanglement and instant communication can only be tested statistically by twin particles demonstrating complementarity in a statistically-significant number of trials. This is similar to ESP and PK (psychokinesis) being tested using a random numbers generator. Yet some physicists have dismissed the ESP/PK statistical experiments as *numerology* even though statistically-significant results have been obtained (see chapter on Physics and Consciousness).

Strong Entanglement of Fermions

Using Stern-Gerlach findings, test whether entangled electrons or other fermions would correlate their behavior. If a magnet applied to one fermion causes a deflection, the twin fermion should show the opposite deflection without passing through a magnetic field. For example if one fermion is deflected upward by the magnet, its twin should be deflected downward, even without a magnetic field.

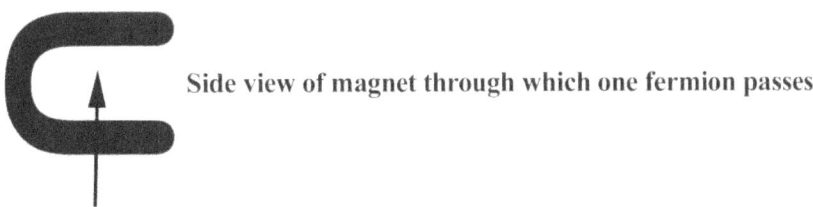

Side view of magnet through which one fermion passes.

Now some physicists might argue that the above experiment using magnets is not what quantum mechanics predicts – it only predicts that when one particle is measured to have a spin direction, the other particle will have the opposite spin direction. However, that does not determine whether the particles had that orientation when entangled locally or whether they acquired that inverse spin direction when measured. To make the distinction between local entanglement and distant effects, one particle has to be changed (after it has been entangled with the other particle) to detect the difference between local determination and distant determination. As we will see, Bell's method of statistically differentiating local determinism from distant determinism is invalid because it makes the wrong assumptions about EPR. Even if one assumes the Copenhagen interpretation that the particle has no definite spin before observed, when observed, the particle changed and took on a definite spin which was simultaneously accompanied by an inverse change in the other particle. If one makes Bohmian assumptions, then the particles already had definite spin, but might have been changed by the measuring device which would be immediately mirrored by the twin particle. Thus, I don't see how this experiment with a magnet would be any different from the experiment with filters and photons. In experiments with photons, the angle of polarization has to be changed after entanglement to see if the other photon will mirror that change inversely. The way of determining if the twin photon has changed is to see if it has the inverse polarization by passing through a filter that would permit this passage.

Perhaps what I am proposing is evidence of strong entanglement rather than weak entanglement.

Popper's experiment measured strong entanglement, i.e., the experimenter changed some the property of interference of one particle, but the other particle did not change inversely. So, if the magnet deflects one particle up, the other particle should be deflected down if strong entanglement is true. To test weak entanglement, a magnet could be placed at the other end to see if the twin would deflect down if its partner deflected up thus indicating opposite polarization.

The following discussion is a response I received from a physicist about my proposed experiment. I have not asked him if I can use his name in my book, so I will leave him anonymous.

Discussion:

Physicist: You said, "So, in Copenhagen, if you change the particles spin from an indefinite state to a spin up, then the other particle should change from its indefinite state to spin down." This is not correct. If you have two entangled spins, and you measure one, then in Copenhagen, this instantly changes the other one to the opposite spin. If, in contrast, you were to take the spin of one and force it to some position, the other one's spin would not change. Perhaps I can clarify this with another example. It is possible (at least in principle) to have a spin-0 state change into a pair of electrons that are going in exactly opposite directions, but the direction of each is not known. Conservation of momentum insists that they must be back-to-back. If you measure one momentum, the other one will be opposite it. However, if you change one of them, for example by passing it through a Stern Gerlach device, then since you have used force on the system, the momenta will no longer be back to back. Pushing on one does not change the other; measuring it does.

My Answer: Again, there is no way to measure a particle without applying a force to one of the particles. As noted, when you send photons through a filter, you don't just passively measure the photon's polarization, you force it to have a polarization it did not already have. In the Copenhagen view, the photon is in an indefinite state of superposition and the filter forces it to have one definite polarization. In the Bohmian idea, the photon already has a polarization – you just don't know what it is. But again, the polarizer may force the particle to change its polarization. In both cases, the particle is forced to change its polarization and its twin is therefore forced to assume the opposite position. The assumption is that the twin particle would be forced by "spooky action at a distance" to assume the opposite position whether there was another filter there to measure it or not. This could be shown by placing the filter of the twin particle a greater distance away from the point of entanglement to see if it assumed the complementary position without a filter to force it. For example, if photon A was forced to a vertical polarization, then photon B could be measured at a distance from the entanglement point greater than particle A's distance at measurement. Photon B should be in the horizontal polarization by the time it reaches that point.

Paraphrasing Physicist: But, as pointed out, when you apply a force to one particle and not the other, you have disturbed the equilibrium of momentum so that one particle has more momentum than the other, and the conservation is lost; therefore, what happens to the particle with greater momentum will not affect the particle with less momentum. The force must be applied to both particles equally as in going through two filters simultaneously. The particles actually act on each other mutually and instantaneously when the same force is applied to both.

> **My Answer:** But as Bohr and Heisenberg pointed out, when particles are entangled, they form one system, so that whatever happens to one particle is immediately conveyed to the other particle of the system, so the force on one would be shared equally with the other even at a distance (recall "spooky action at a distance"). When Einstein objected to the idea of a faster-than-light signal between entangled particles at a distance, Bohr said that the two particles make up one system and cannot be individualized after entanglement. This statement indicates that the force on one particle would be applied equally to the other particle even at a distance so that conservation of momentum is preserved. This is the troubling aspect of the mysticism that is inherent quantum physics – particles are said not to behave in a logical way. This argument of force vs. measurement seems like a hedge and an insulator from falsification of entanglement theory. The only way to measure a quantum particle is to apply some force to it because it cannot be seen otherwise – there is not passive measurement of quantum particles.
> **Physicist:** In a quantum entanglement experiment, you get one measurement on the particles. After that one measurement, decoherence or wave collapse sets in, and the entanglement is destroyed.

Weak Entanglement Using Stern Gerlach-Gerlach Device

Even if the proposed experiment on strong entanglement is rejected, it seems to me that the Stern Gerlach Device provides a better instrument for testing distant communication between entangled particle than other methods I have seen. If entangled pairs of electrons are shot through separate Stern-Gerlach Devices on opposite ends of the experimental set-up, then if one particle is deflected up, the other should be deflected down. QM would predict that 100% of the electron pairs would be deflected in the opposite direction because they are in superposition, carrying all possible polarizations before measured. EPR would predict that perhaps 50% would show opposite polarization. This would be true of EPR because about 50% of the electrons will pass through without being deflected because their polarization is right to left rather than vertical. However, QM assumes that the pair has all possible orientations and should respond to the up and down detector because the electrons are carrying both sets of information – up and down and side to side; whereas EPR says the electrons are already set locally in one orientation or the other. Chance expectation would be around 25% because the particles would not be entangled and, half the time, they would pass through because of horizontal polarization and half the time they would be deflected on the vertical axis. Of the half that are deflected vertically, half of those (25%) would be deflected in the same direction and half (25%) in the opposite direction. If an experimenter wants to further complicate the experiment to test for stronger entanglement, she could run the electrons through a second Stern-Gerlach Device to see if those electrons that pass straight through because of a side-to-side polarization, would show opposite polarization or same polarization. If those passing straight through showed more opposite polarization than same polarization, that would be additional evidence for entanglement.

Weak Entanglement of Fermions

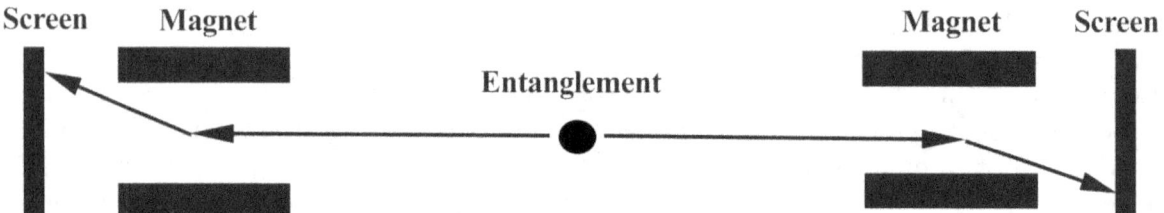

Using Stern-Gerlach Device, test whether entangled electrons or other fermions would correlate their behavior. If a magnet applied to one fermion causes a deflection, the twin fermion should show the opposite deflection. For example if one fermion is deflected upward by the magnet, its twin should be deflected downward.

Sequential vs. Simultaneous photon arrival at filters: If, say, the 0/180 filter were placed closer to the light source than the 90/270 filter, the photons hitting the 0/180 filter, would be polarized on that axis, thus causing the photons traveling toward the 90/270 filter to be polarized in the 0/180 axis which would mean that they would be blocked by the 90/270 filter. However, this experimental set up (with filters at different distances from light source) would not support instantaneous communication but would support non-local communication. As indicated, to show instant communication, the filters have to be equidistant from the light source.

Moreover, if filters are changed randomly while the particle is in flight by a switching mechanism, then any matching cannot be attributed to local, hidden variables created at the time of entanglement – unless one believes, as Wheeler did, that information can go backward in time.

Testing Bell's Inequalities Using Spin, rather than Polarization, Yields Different Results

Bell's theorem is a proof by contradiction. Making EPR assumption of local, hidden variables allegedly leads to a reductio ad absurdum. The mathematical inequality between the quantum prediction and the EPR prediction indicates that inverse correlation of particles could not be the result of local entanglement. Research results supposedly shows that local, hidden variables predicted by EPR can be ruled out. Here is my analysis based upon my reading of the interpreters of Bell.

Up to this point, we have been discussing Bell's Theorem using polarization of photons which are classified as bosons which are assumed to be massless. Now let's look at the Bell's test using a fermion such as the electron which has a charge and acts as a tiny magnet with polarization. First let's view the predictions and assumptions of the EPR interpretation and the quantum interpretation (which is basically synonymous with the Copenhagen interpretation since that is the dominant paradigm in modern physics).

The assumptions of Bell (or perhaps Bell's interpreters of his theorem) are supposed to be EPR assumptions which can be reduced to absurdity, but they are not EPR assumptions. They are quantum and random assumptions.

1>**EPR assumptions/predictions**: When particles are created or entangled, they immediately have opposite spin 100% of the time which is necessary to preserve angular momentum. The angle of spin (somewhere between 0 and 360) is set at the time of entanglement and embedded in the twin particles, so that even if they are separated, this information remains in the particles. If one particle is spin up, the other will be spin down in the same axis such as 0/180 – not in any combination of lines. There is no half-down, ¾ down correlation - there is only fully down (180-degrees apart) correlation. The whole point is to conserve angular momentum, and momentum cannot be conserved half-way or ¾ of the way. Given this interpretation, all particle pairs will have opposite and equal spin, that is, they will always be measured to have different states (A up/B down or A down/B up) and will never show the same states (A up/B up, or A down/B down). If particles show the same states, then they were not entangled to begin with.

2>**Quantum (Copenhagen) assumptions/predictions:** When particles are entangled, they acquire some basic information, that is, to always have opposite spin when measured, but the direction or angle of spin is undetermined until measured. The superposition interpretation is that the particles have all possible angles of spin until measured – the other interpretation is that they have no definite direction of spin before measured. So, in Q-M, half the information (particles will have opposite spin) is embedded locally, but the other half (actual angles of spin) is not. Since the constraint to be measured with opposite spin, even in the quantum interpretation, is built into the entanglement process to conserve angular momentum, then the prediction should be the same as EPR's prediction – that 100% of the time the entangled particles will show opposite spin to conserve angular momentum when measured. If the pair shows the same angle of spin, then they were not entangled. However, the interpreters of Bell tell us that the odds for twin particles are 50% for same spin and 50% for opposite spin. Of course, entanglement means particles will have opposite spin or different spin when measured at a distance. Is there any other meaning to the word "entanglement?" If spin is not entangled, then what is? Entanglement means that particles have mirror-correlated DNA, but the exact complementary angles are determined at measurement.

3>**Random pairing of particles without entanglement:** The real assumption of the interpreters of Bell are that the particles are independent of each other so that sometimes particle A can be spin up and particle B can be spin up also and vice versa. If this same behavior occurs, then there is no entanglement and the odds would be 50/50 of getting a *same* response or getting a *different* response. After all, the experiment boils down to the fact that either the pair is opposite in spin or the same in spin. There are two ways of getting same spin (particle A up and particle B up or, vice versa, particle A is down and particle B is down). And, there are two ways of getting a different measurement (particle A up, particle B down or particle A down, particle B up). So, regardless of the number of detectors, it boils down to a 50/50 chance of getting the same or different response. *As can be seen, the Random pairing of particles gives the same prediction as the quantum (Copenhagen) prediction – 50/50 either way.* Since the quantum interpretation does not differentiate itself from the random interpretation, it is not a scientific theory because one could attribute a 50/50 result to either interpretation. There is an equality here - not an inequality. The inequality exists between the EPR interpretation and the random/quantum interpretations. *The results for entanglement cannot logically be the same as the results for non-entanglement.*

The fact that quantum theory does not beat the chance expectation of 50/50 shows that it does not really predict anything that a flip of a coin would predict. Most theories predict that the results will be a deviation from chance. To sum it up, this experiment does not differentiate between EPR and quantum interpretations.

The true test of this theory could be done with two Stern-Gerlach devices. Entangled particles would be shot in opposite directions to go through identical Stern-Gerlach devices at equal distances from the point of particle creation. If entanglement is true, then the particles would be seen to have opposite spin and opposite angles of spin (particles would **not** be limited to 3 angles of spin). Of course, this result would not tell the experimenter whether this inverse correlation of particles was due to local entanglement or communication at a distance. Probably, the only way to separate local entanglement from distant communication is to change one of the particles in flight to see if the other follows suit. Now some theorists claim that to change a particle after it has been entangled is to create decoherence and destroy the entanglement. Others say this is not the case – that entanglement is stronger than that. Actually, using the experimental set-up mentioned above with three angle detectors, the experimenter is forcing particles to take only one of the orientations in its superposition. Still others say that if you intentionally force one of the particles to take on a state it didn't have, then you destroy the entanglement which puts consciousness back in the driver's seat as in Copenhagen. Yours truly does not see the difference between forcing a particle to change through measurement and forcing a particle to change in any other way. To test for strong entanglement, then, the Stern Gerlach devices could be used to determine if changing one particle will instantly change the other. If it is possible to deliberately change a particle and have its behavior mirrored by the other, then instant communication at a distance would be possible.

The experiment to test Bell's theorem is set up so that there is a device for entangling electrons and sending them in opposite directions to different detectors. Each detector has three settings for testing the angle of spin for each particle. These angles are 0-degrees, 120-degrees and 240-degrees; however, since particles can have opposite angles of spin, there are actually six different detectable angles 180-degrees apart (0, 60, 120, 180, 240, 300, 360), so that only if particles have these orientations will they be detected. If a particle does not have one of these orientations, it would have been forced and that would destroy entanglement. The spin-angle of the detectors on each device is selected randomly. The prediction of both theories is that the particles will always have opposite spins, however, quantum theory says the angle of spin will be determined at the time of measurement.

Without going into the mathematical details, if both particles pass or if both particles do not pass, their spins are correlated with the angle of the detector and this is counted as a match or *same*. If one particle passed and the other did not, that means that the passing particle was oriented the same as the detector and the other particle was not – this would be a mismatch or *different*. Given these assumptions, EPR (that there is a pre-existing angle of spin) predicts that 55% or more of the particle-pairs will match, that is, one will be up and the other down. Experiment shows that only 50% do not match. Hence the inequality between prediction and experimental result supposedly rules out EPR's claim that the particles had a definite angle of spin acquired at the time of entanglement and supports the quantum idea that the particles acquired their correlated angle of spin at the time of measurement at a distance from each other. But does this experiment really force that conclusion or is the experiment a self-fulfilling prophecy created by the original assumptions.

Analysis of the Bell experiment using spin orientation
1>Bell was using quantum assumptions to rule out EPR – he was not using EPR assumptions as advertised. Bell's assumption was that each pair of particles, true to Copenhagen, was carrying information on all possible spin orientations the particles could be in. For example, there are eight combinations of information that could have been carried by each particle, and its twin particle would have the opposite spin (if one is up the other is down). To take Copenhagen to the extreme, each particle pair would be carrying information on all eight combinations, but Bell assumes they are carrying only a set of three at a time.

	Particle A		
	Angle 0/180	**Angle 120/300**	**Angle 240/60**
1	UP	Up	Up
2	Up	UP	Down
3	UP	Down	UP
4	Up	Down	Down
5	Down	Down	Down
6	Down	Down	Up
7	Down	UP	Down
8	Down	UP	UP

*Particle B would have the opposite spin orientation if entangled with A.

Using number 3 combination (highlighted in gray above), if the particle A had "up" information in the 0-angle, particle B would have down orientation in the 180-angle. And the same would apply to 120/300 and 240/60 angles since the entangled particles allegedly always have opposite spin. Bell is interpreted as saying that each particle carries all three orientations. However, this is not what EPR would predict. EPR would predict that only one of these orientations was established at entanglement and that the particles would be in that orientation upon arrival at the detector. The particle would not have any information on the other two angle combination. I think Bohmian mechanics, which makes many of the same assumptions as EPR would say the same – the particles have one definite orientation and no other information about other orientations. Hence Bell commits an error right out of the gate in applying Copenhagen, quantum assumptions to EPR (and probably Bohmian Mechanics as well). Using EPR's own assumptions, the odds would **not** be 45% for getting a matching response (same), the odds would be 1 in 3 or 33% of both particles passing creating a match. It would be difficult to determine the odds of getting a match by both particles in the same position – since entangled particles could have had any orientation. Actually, it would be hard to determine the odds if there are only three options if the experimenters count a match to be both particles do not pass.

What we have in this experiment is that there are two entangled particles that have opposite spin in any conceivable direction, and two detectors, each of which have three settings for detecting specific spin angles. The settings of each machine are randomly and independently selected so that there are nine combinations of settings. Now, if we assume that entangled particles have only three possible orientations (a wrong assumption), we might estimate what percentage of twin particles will match their behavior by both passing through detectors or both being blocked or not matching their behavior by one passing through and the other not passing through.

Detector A	Detector B		
	0/180	120/300	240/60
0/180	Pair 0/180 detected		
120/300		Pair 120/300 detected	
240/60			Pair 240/60 detected

If the particle-pair had only three possible orientations and only one actual orientation at a time (180-degress apart), then there would be a 33% chance of any given pair with opposite spin passing both detectors because the pair would only pass if both detectors matched the exact orientation of the particles. If the particle pair passed with a non-matching orientation, then the orientation of the particles would have been changed and according to quantum theory, such change would have destroyed the entanglement through decoherence. Now if one allows the twins to have any orientation (rather than just three), then the same result occurs, that is, if a particle passes through a detector with a different orientation from the particle, then the particle will have its orientation forced and therefore decoherence sets in and the entanglement is destroyed.

It is interesting that Bell used Copenhagen assumptions to design his thought experiment, but thought his theorem supported Bohmian Mechanics. Here's what Bell had to say about Bohm's ideas:

But Bell supported pilot-wave theory. He was the one who pointed out the flaws in von Neumann's original proof. And in 1986 he wrote that pilot-wave theory "seems to me so natural and simple, to resolve the wave-particle dilemma in such a clear and ordinary way, that it is a great mystery to me that it was so generally ignored (Wolchover 2014)."

Now, how would Pilot Wave theory solve the "spooky action at a distance", faster-than-light communication conundrum? I suppose Bohm might say that that the entangled particles are connected by one pilot wave that directs their paths and behaviors at a distance. As to whether information could be communicated within the pilot wave instantly is a matter of conjecture. However, unlike Copenhagen, the particle would have only one state at a time and there would be no need for superposition with all possible states. Any undetermined states of the particle would be due to our ignorance, not the particle's ignorance or lack of certainty. It is interesting to note that there is an entourage of particles that surrounds the electron. The evidence for this comes from particle colliders and the fact that when electrons are sent through a wire to form a current, they create a magnetic field that can be detected by a compass. Perhaps these companion particles carry the hidden, local information about the electron and are a part of the pilot wave.
But then, it is doubtful they could travel faster than light.

Fallacies of Bell-Aspect Experiment and Proposal for a Different Method

Again, the fallacies in Bell's Inequalities boil down to language and logic - the theme of this book. The Bell Theorem as presented by the interpreters of Bell involve several semantic and logical fallacies. Let me count the ways.
1>Bell said that variables that account for entanglement cannot be local, and measurement at a distance from the point of entanglement accounts for mirror-matching of particles. Yet, Bell and Q-M assume that at the time of entanglement, the particles acquired opposite spin - the only thing that was not determined at entanglement in QM was the angle or axis of spin.
2>Bell, or Bell's interpreters, try to reduce EPR to absurdity by making EPR assumptions and then showing that they cannot be true. However, Bell, or interpreters, make QM (Copenhagen) assumptions instead. To wit, EPR assumes that entangled particles will acquire complementary properties when entangled such as a definite spin and definite axis (for example, if one particle is spin up, the other will be spin down at 180-degrees apart). However, Bell's interpreters indicate that entangled particles have three plans or three sets of information that correspond to the three axes (0, 120 and 240). This assumption is more like *superposition* in which particles are carrying all possible spins. Again, EPR indicates that the particles have one spin and one axis acquired at the point of entanglement.
3>Bell assumes that a particle pair can be measured to have half or ¾ entanglement rather than whole entanglement when Q-M calls for whole-integer increments in most cases. If entangled, fermions (electrons) are supposed to be 180-degrees from each other. If electron A is measured at 0-degrees, electron B should be measured at 180-degrees, and if B is measured at some other angle, then it is either not entangled, or it has been forced into a specific axis. Some sources

indicate that if a particle is forced into a particular axis, decoherence occurs and entanglement is destroyed.

4> The predictions of Q-M are the same predictions as the prediction derived from independence and chance. The prediction of Q-M is that there is a 50/50 chance of getting a "same" behavior or a "different" behavior from entangled particles. Since obviously twin particles will behave the same (both up or both down) or different (A up, B down or A down, B up), then in a large number of trials based on chance, the flip of the coin will yield 50/50 results if either QM or independence/chance is true. Therefore, there is no distinction between QM predictions and chance.

5> Bell's experiment is supposed to make EPR assumptions and show they cannot be true by *reduction ad absurdum*, but the assumptions made are quantum assumptions, not EPR assumptions.

6> Bell's interpreters give different statistics for electrons and photons for violating Bell's inequalities. Using electrons (fermions), the EPR predicted odds are 5/9ths (55%) for getting different responses and the predicted odds by QM are 50/50. Using photons (bosons), the EPR prediction is 33%, and the QM prediction is 25%.

7> Confusion about spin up and spin down. With electrons, for angular momentum to be conserved, the axes of spin must be opposite each other (180-degrees apart). Therefore, if particle A is at 0-degrees, then particle B cannot be at 120-degrees. B must be at 180-degrees to completely conserve momentum. Also, with photons, if particle A is at 0-degrees, then particle B must be at 90-degrees, not 120-degrees. To force the photon from 90 to 120 degrees is to create decoherence and destroy entanglement according to some physicists. Forcing particles into different spins would throw the statistics completely off. However, some physicists claim that photons (bosons) are polarized at 180-degrees just like electrons. Here again we see confusion in the concept of entanglement.

8> Contradiction on angle of spin. If particle A is at 0-degrees up, B should be at 180-degrees down, but if it encounters a 120 filter, some say that it has a 50/50 chance of being either spin up or spin down, but it is more likely to be spin up. Now, to begin with, if particle A is spin up at 0, its partner will be opposite or spin down at 180, and since 120 is closer to 180, B would be more likely to be spin down. Also, if there is a 50/50 chance of B being up or down at 120, then it cannot be more likely to be up rather than down. If B is more likely to be spin up, then the odds are not 50/50. Furthermore, these are quantum assumptions, not EPR assumptions, and the test is supposed to use EPR assumptions to demonstrate that they cannot be true. Quantum assumptions are based on superposition (each particle has all possible spins) rather than EPR assumptions - that each particle has one definite spin axis acquired at the point of entanglement.

Consider the following table which shows that with 0, 120, 240 detector angles randomly selected, there is never a combination of detectors which will produce an alignment 180-degrees apart. Thus, any spin up/spin down measurement with these detectors will be forced and therefore destroy entanglement. Therefore, this test of entanglement and Bell's inequalities is flawed from the outset.

QUANTUM QUANDARIES AND QUACKERY — Various Interpretations of QM

9 Random Detector Combinations			
Detector A	Detector B		
	0	120	240
0	0/0	0/120	0/240
120	120/0	120/120	120/240
240	240/0	240/120	240/240

The compass below shows six paired angle combinations regarding fermions such as electrons. Each pair is 180-degrees apart.

1> EPR assumptions: particles obtained opposite spin locally at entanglement.

2>Quantum assumptions: particles acquired potential opposite spins at entanglement but manifested a definite angle of spin only when measured at a distance.

Therefore, if A is spin up at 0, then B must be spin down at 180 in either EPR or QM. B cannot be down or up at 120 and conserve momentum in either EPR or QM

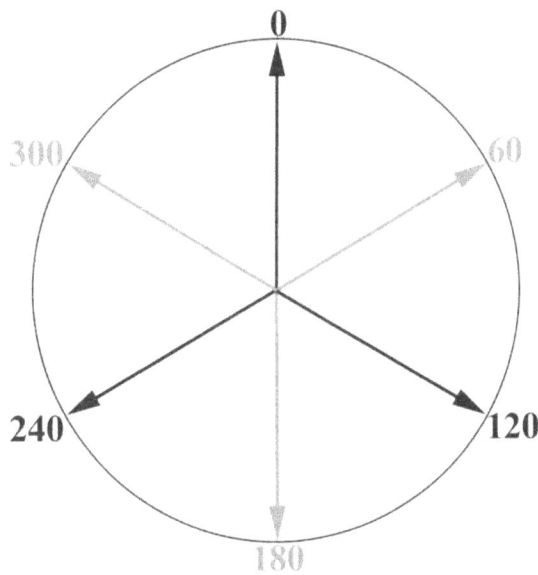

Where Information Between Entangled Particles is Acquired		
	At Entanglement	At Distance
EPR (Einstein, Podolsky, Rosen)	*Opposite Spin (1 up, 1 down) *Opposite Angle (Always 180 apart to conserve angular momentum)[1]	No new info acquired at distance
QM (Copenhagen)	*Opposite Spin (1 up, 1 down) *Superposition of Spin Angles[2]	Angle of Spin (0/180, 120/300, 240/60) faster than light speed[3]

Bohmian Mechanics	*Opposite Spin (1 up, 1 down) *Opposite Angle (Always 180 apart – same as EPR)	Same as QM, Bohm implies that particles can change angle of spin at a distance.[4]

[1] In EPR, twin particles are believed to acquire a definite spin and angle of spin at entanglement. Bellian interpreters assume that the particles have three plans, one for each angle of measurement (0, 120, 240); however, this assumption is not an EPR prediction and therefore nullifies the theorem and the experimental results obtained from QM assumptions.

[2] For particles to be entangled at their birth means that they had to share some kind of information – otherwise entanglement has no meaning. In QM, the information is that twin particles will always be 180-degrees apart to conserve angular momentum; however, the exact angle of spin is not determined at entanglement. Instead the particles are said to be in **superposition**, that is, they are in every possible position (0, 120, 240, etc.) until a measurement is made, thus collapsing the wave to actualize one definite position. Instead of all possible positions, some physicists say the particles have no definite position until measured.

[3] EPR indicates that no information can be transmitted faster than light, but Bell's theorem indicates that faster-than-light transmission is possible if particles are part of one system, i.e., entangled. Also, some theorists argue that if some property of the particles is forced to change (e.g. angle of spin), entanglement is destroyed, and the particles will not coordinate their behavior. However, that assertion contradicts the "action at a distance" argument of QM. If particles are compelled to change from a superposition state prior to measurement to a definite, single state after measurement, then the particles have been forced to change. Also, in polarization experiments with light, the angle of polarization is forced to change in these experiments and coordination of behavior between the particles is allegedly not lost. Other physicists get around this problem by saying that the particles get one correlated change and then decoherence sets in. That would mean that if particle A is measured at 0-degrees, particle B would snap to 180-degrees and that one change would destroy entanglement and decoherence would occur.

[4] Bellians start with QM assumptions while indicating they are making EPR assumptions, and then Bell says that his theorem and experiments support Bohmian Mechanics which indicates that the twin particles have a definite spin and angle at entanglement (locally) and can also communicate at a distance (supposedly faster than light). This assertion would also indicate that particles *can indeed be forced to change* at a distance and still mirror each other's behavior – thus retaining their entanglement for one change.

The following charts represent EPR and QM predictions for various combinations of detectors selected at random.

QUANTUM QUANDARIES AND QUACKERY
Various Interpretations of QM

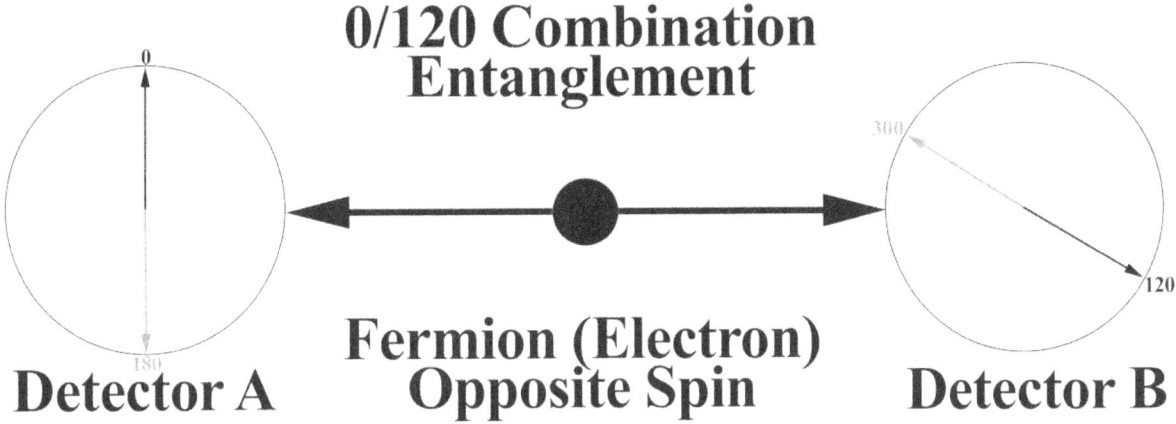

In this combination, the detectors are not aligned with the angle of spin of the particles whether making EPR or QM assumptions. In both EPR and QM, the particles must be aligned 180-degrees apart – whether made that way at entanglement or when measured. Therefore, if particle A is measured spin up at 0, then B should **not** be measured spin down by a 120 or 300 detector since it would be misaligned with the B's orientation. If it is measured at 120 or 300, then either the particles were not entangled to begin with, or one particle was forced out of its alignment and the entanglement is destroyed. Or vice versa, if B is measured down at 120, then A will be at 300, and should not be measured at 0 or 180.

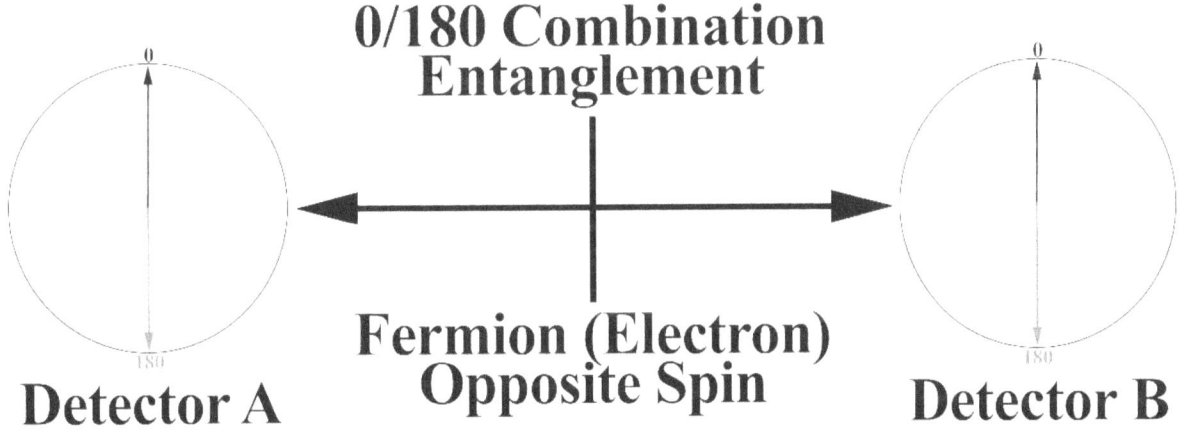

In this combination, if both detectors are in 0 spin-up orientations, the detectors would be misaligned with the particle orientation and entangled particles would not be measured at opposite and equal spin. Therefore, the result would be the same whether making EPR or QM assumptions.

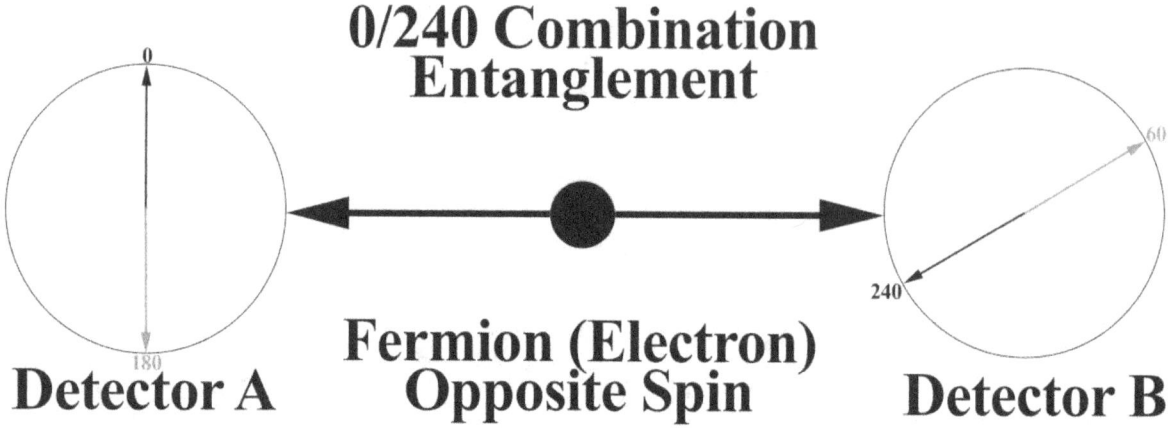

Once again, the detectors are misaligned with any possible spin angle of entangled particles, and if A is measured spin up at 0, then B should be measure spin down at 180. If B is measured at 240 or 60, then either B is not entangled with A, or B has been forced into that axis which destroys entanglement.

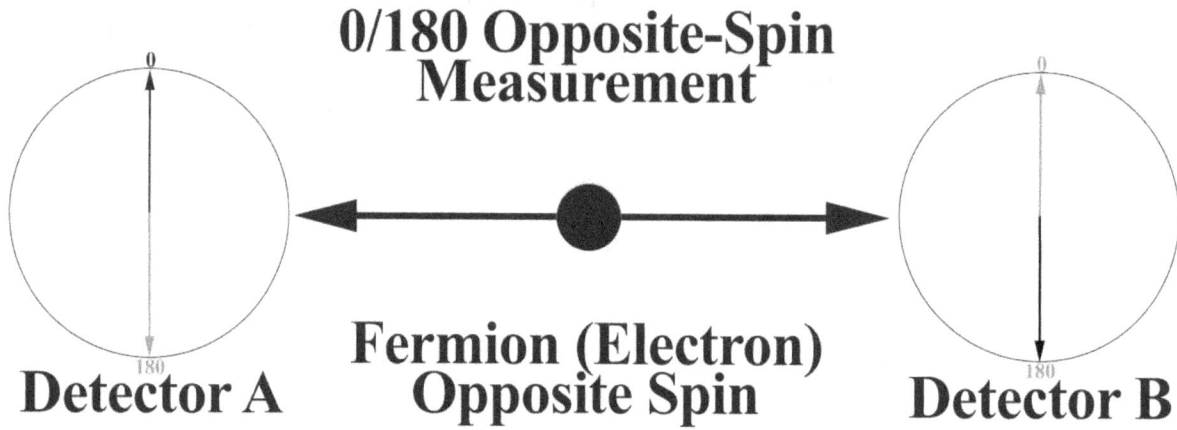

In this combination, using QM assumptions, both particles would be detected since the detectors are aligned 180-degrees apart. This would be true, since according to superposition, the entangled particles are aligned 180 apart on every possible axis before measurement, therefore both would be detected with any combination of detectors that are 180-degrees apart. However, with EPR assumptions, the twin particles would have only one specific orientation 180-degrees apart at the time of entanglement – they would not be in every conceivable position when measured. Thus, if there were a detector for every one of the 360-degrees of the compass, there would be a 2/360 (1/180) possibility of both particles being detected 180-degrees apart if EPR is true. If QM is true, then both particles would be detected at 180-degrees apart 100% of the time since particles carry all possible combinations. This method is a much clearer and direct way of comparing EPR and QM predictions than the Bell method as will be shown.

Conclusion: Because of the many erroneous assumptions in the Bell method noted above, it is not a fair test of EPR. Using detectors with six angles when detectors are always aligned 180 degrees apart, QM would predict 100 % detection and EPR would predict that 2/360 or 1/180th of the time, both particles would be detected on those angles. Of course, the detectors would have to be calibrated to 1-degree accuracy, so that if a particle is at ½ or other fraction of a degree between one number and the next, the particle would be detected. This method would be clear-cut and eliminate the confusion one sees among different interpreters of Bell. Furthermore, Bell embraced Bohmian Mechanics which contradicts QM. QM indicates that particles have no definite angle of spin when entangled, Bohm, like EPR, says that they do have a definite angle of spin when entangled – it's just that the angle is unknown until measured. However, Bohm, like QM, but unlike EPR, says that particles can talk instantly at a distance. Although Bohmian predictions are the same as QM predictions in the end, Bell's theorem supposedly rules out local, hidden variables so Bohm's idea definitely contradicts Bell's original assumptions. The following table shows the assumptions made by EPR, QM and Bohmian mechanics.

	Orientation of Particles set Locally at Entanglement	Orientation of Particles set at Distance
QM (Copenhagen)	No	Yes
EPR	Yes	No
Bohmian Mechanics	Yes	Yes

Proposed Experiment to Test EPR and QM

Since Bell's inequalities makes incorrect assumptions about EPR and uses QM assumptions to rule out EPR. I propose the following experiment. The basic experimental set up would be that detectors will always be locked at 180-degrees apart, but they can be randomly spun to any of the 360-degree angles possible (sort of like the Wheel of Fortune of TV fame). If detectors are locked but spun randomly in this manner, my prediction is that, if EPR assumptions are true, then detection of the spin angle of entangled particles will be rare, but if QM assumptions are true, the detection of the spin angle of entangled particles will be common (theoretically 100%). The following are my assumptions regarding such an experiment. See if you agree with my assumptions and interpretation of Quantum Theory.

1>Entanglement means that when two particles are brought close together or are emitted together, they become linked by one wave function that controls both particles. In order to satisfy conservation laws, entangled particles must have opposite, complementary states so that they balance to zero, that is, if one is spin up, the other must be spin down 180-degrees apart. So, spin at +180 minus spin at -180 = 0. RIGHT OR WRONG?
2>The EPR thought experiment indicates that these opposite, complementary states of entangled particles were acquired LOCALLY at the point of entanglement, but the entanglement is HIDDEN until a measurement is made. RIGHT OR WRONG?
~Therefore, if one sets the spin-detection apparatus 180-degrees apart and spins it randomly always maintaining the 180 linkage, it will be rare for the apparatus to match the orientation of the particles if EPR is true. For example, if the detector is randomly set to measure at 0/180 then the particle must have been entangled at 0/180 for the detector to measure them. Assuming there is one-degree of tolerance, the odds are 1 in 180 that the detectors will match the spin orientation of the particles. The probability is 1/180 rather 1/360 because there are two ways that entangled

particles can match a 180-degree set. If Particle A is at, say, 0, then Particle B would be at 180, or conversely, if Particle A is at 180, then and Particle B would be at 0. RIGHT OR WRONG?
3>The Copenhagen interpretation (dominant Quantum interpretation) says that the particles do not have one definite spin orientation at entanglement – they have every possible spin orientation until measured, and measurement collapses the wave to fix one particular spin orientation. However, the particles, if entangled, will manifest 180-degrees apart when measured. These multiple states of a particle are known as SUPERPOSITION. RIGHT OR WRONG?
~Therefore, no matter what the position of the detectors (as long as they are kept 180 degrees apart), the particles will always be detected since they had every possible spin before detection. RIGHT OR WRONG?
~Therefore, if Copenhagen is correct, the particles will be detected 180-degrees apart 100% of the time because no matter what position the detectors are in at 180-degrees apart, one of the multiple states of the particles will match it. RIGHT OR WRONG?
~If EPR is correct and the particles were fixed locally at 180-degrees apart, then the particles will only match detectors if the detectors are in the same orientation as the particles when entangled which would be about $1/180^{th}$ of the time. RIGHT OR WRONG?

Please consider the following experimental set up to test QM vs. EPR assumptions.

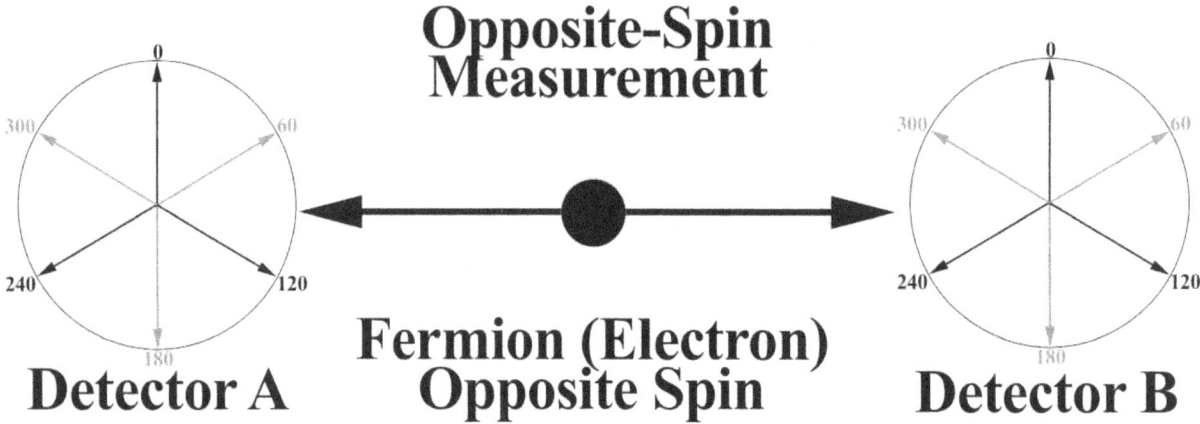

The above graphic illustrates a proposed experiment that would be more clear-cut and definitive than the Bell experiment I believe.
1>Shoot entangled electron pairs at detectors at opposite ends of the device.
2>If QM is true, 100% of the pairs should be detected at 180-degrees from each other because of *superposition*, i.e., the twin particles are carrying all possible angles of spin at 180-degrees apart.
3>If EPR is true, then the angle of spin will always be 180-degrees apart, but could be at any angle which was determined at time of entanglement. Therefore, EPR would predict that the particles will be detected only if the detectors are in the same alignment as the alignment of the particles when locally entangled. Thus, if the particles are entangled at 0/180, they will be detected only if the detectors are set at 0/180. Assuming that there is one degree of tolerance in

the detection, that would mean that in any given trial, there would be a 1/180 chance of both particles being detected according to EPR. The odds would be 1/180 rather than 1/360 because there would be a duplication of detections with two particles at 180 apart. For example, there would be a detection is Particle A was at 0 and B at 180 or if B was at 0 and A was at 180. For each degree there would be this inverse duplication.

4>If there is *no* entanglement, then individual particles (in another set-up) could be detected that were not spinning180-degrees from each other. For example, one particle might be at 0 and the other at 120 and both might be detected (even though they are out of the 180 alignment) because they are spinning independently of each other. Since the particles are not necessarily spinning at opposite angles, up and down relative to each other, they are not entangled because angular momentum would not be conserved.

Other theories and findings in physics contradicts QM (Copenhagen) assumptions such as the Heisenberg Uncertainty principle of the Copenhagen interpretation which indicates that the researcher cannot know opposite, complementary properties of particles at the same time. Not only can one not know these complementary properties, but they do not have definite properties until measured in QM speak. Here are the theories which contradict uncertainty until measured.

1>*In the quantum foam*, entangled pairs of matter/anti-matter particles are theoreticaly born from the vacuum. These particles are said to have opposite spin, opposite charge, opposite time, etc. at birth when entangled. Since this quantum bubbling is said to be happening everywhere, there are no known observers in intergalactic space to observe and measure quantum particles when they bubble out of virtually nothing. Therefore, the particles must have had *definite* properties when born into a state of entanglement as EPR and Bohm assert.

2>*Pauli's exclusion principle* indicates that electrons with the same properties (e.g., same spin) cannot occupy the same locus in the orbitals of the atom. Electrons must have opposite, complementary properties (such as opposite spin) to occupy the same locus. Since electrons are constantly being ripped from their orbitals and reabsorbed, there can be no ubiquitous human observers making their measurements to compel electrons to take on a definite state when this process happens, so the electrons must have had definite opposite spin already before getting paired up in an orbital. Both these concepts, then, supports the EPR concept that particles are hatched and entangled with definite opposite, complementary properties – whether they are measured or not. To assume that human measurement is required for the micro world to operate is quite anthropocentric. How did particles do their thing all those millennia before humans evolved on this planet. Wait, wait, don't tell me – when humans were born their observations went backward in time to create the universe we see. Really??

Entangled photons subjected to different environmental conditions

The following thought experiment is similar to Popper's experiment in which entangled particles are placed in different situations to see if they will correlate their behavior.

1. Entangle photons
2. Send one through atmosphere so that it slows to .75C.
3. Send the other through the virtual vacuum of space so that it should retain maximum speed of C.

What would be the outcome of such an experiment? Here are the possibilities.
1. Space particle slows to .75C or particle in atmosphere speeds to C.
2. Since the particles are traveling at different speeds, would instantaneous communication be preserved, or would decoherence occur?

More Loopholes to overcome in Testing Bell's Inequalities: There are many more loopholes to overcome than the one posited, most of which come from the more exotic theories of modern physics dealing with the human factor in determining the outcome of an experiment. To begin with, if the experimenter plays a role in collapsing waves and determining the outcome of an experiment, this is a confounding effect that will be difficult to overcome since it is hard to imagine an experiment without an experimenter. The suggestion that light from distant quasars arriving to Earth at an even - or odd - numbered microseconds could be used to eliminate any experimenter effect is potentially problematic. Not the least problem is that all particles were theoretically entangled at the beginning of the Big Bang in the Bohmian interpretation, and unless particles can become untangled by decoherence, they will forever be able to communicate instantaneously. That might be a good problem to have, if the theory is true, because previously entangled particles would guarantee the result for the two current particles that are being tested for correlation. However, if this outlandish idea turned out to be true, perhaps Bohm would be vindicated in his idea that all particles throughout the universe are entangled in a seamless whole. Perchance that would not give evidence for the quantum interpretation the researchers are looking for, since Bohm's theory is not widely, regarded as tenable.

Then, of course, you have the participatory anthropic principle, championed by John Wheeler. If his idea is true that the universe doesn't come into being until we observe it, then the experimenter's thoughts could go back in time and create the very photons that would guarantee the experimental results desired. Then there is Bohm's disciple, Aharonov, who says that in the Copenhagen interpretation, the cat is neither dead nor alive until a probability wave coming back from the future meets and collapses the wave of the present, so that the vital signs of the cat are found or not found. If there is a wave coming from the future (perhaps from one's future self), then theoretical bias of one's future self could contaminate the experiment.

If these theories sound far-fetched, I didn't make this stuff up. Someone please tell me if I have misconstrued any of these theories that seem to come from the far side of the Twilight Zone. It would appear that it is nearly impossible to close all these loopholes courtesy of the exotic theories of modern physics. Another problem I have with the experiment is that physicists are trying to test the correlation between two photons. Given the infinitesimally-small size of the photon (so small that no microscope can come close to seeing it), how does the experimenter know that s/he is releasing only one particle at a time. This level of claimed precision arouses much skepticism in this author. I would think that anytime you release light, you are releasing a multitudinous stream of photons. Secondly, what percentage of hits or correlations would constitute a successful result? If the percentage is less than 100, then we are still talking about quantum statistical probabilities again. Such probabilities have been calculated according to Powell (2016) who cites loop-hole free experiments in Singapore and the Netherlands: "The chance that the strongest of these results could have happened without entanglement is less than 1 in 4 million (p. 60)." Of course, that statement doesn't make it clear whether the correlation between particles was created locally by entanglement or non-locally.

Strong Entanglement vs. Weak Entanglement

Entangled particles seemed to pass the Aspect test of instant communication in matching their polarization inversely. However, particles that are entangled seemed to have failed Popper's test of strong entanglement which involved passing entangled particles through different size slits to see if they would show matching interference patterns. I would like to propose a simplified test which would involve strong entanglement. Using known probabilities of unentangled photons passing through various combinations of polarizers at different angles, the experiment would involve determining if the entangled photons would follow the same probabilities if passed through the same combination of filters. Aspect's entanglement experiment involved passing photons through one filter on each end of the arms of the apparatus. A stronger test would be to see if the photons would match each other's behavior if they pass through more than one filter. For example, if single photons pass through a 0-degree filter then through a 45-degree filter, 50% of the photons will make it through both polarizers. The experiment would involve sending one of the entangled pair through a 0-degree and a 45-degree filter to see if the other photon would pass through a 45-degree filter without first passing through the 0-degree filer. The hypothesis is that if one particle is polarized in the vertical dimension (0-degrees), its twin particle will also be polarized vertically (0-degrees) although in the opposite direction. Then each particle will have a 50-50 chance of passing through the 45-degree filter if strongly entangled and if the 45-degree filter is placed equidistant from the center of the apparatus. The following table shows various combinations of filters with known probabilities of passing. Other combinations with known probabilities of passage could be used as well. If experimental results for entangled particles showed similar probabilities to individual unentangled particles in passage rates in a large number of trials, then entanglement would have a stronger case.

Nutshell: In the above proposed experiment, if the twin particles are entangled, they will manifest the same percentages of passing on both arms of the apparatus, e.g., if 50% of the particles pass through the filters on the left, then 50% will pass through the filter on the right.

> If the particles are not entangled, the particles on the left will conform to the percentages of passing as indicated, but the particles on the right will show near 100% passage.

There are some apparent quantum anomalies in the percentage probabilities in the table. If one combines a 0-degree with a 45-degree filter, the passage rate is only 50%. However, if one interposes a 22.5 degree (1/2 of a 45-degree) filter between the 0 and the 45, the passage rate goes to 70% or 20% more than the 0 and 45 combo alone. Perhaps this is no spooky quantum mystery if one considers that it is easier to change the polarization of photons in small increments rather than large increments. So, a 0-45 combination is halfway to completely blocking the light which occurs if contiguous polarizers are turned at 90-degree angles to each other such as 0-90. A 22.5 filter is only ¼ of the way to total blockage. If one continues to place smaller increment filters (1/8, 1/16, etc. in the lineup between 0 and 45, there is more and more light that gets through. So, by coaxing the photons to change polarization by smaller and smaller amounts, the photons can make the change before being blocked. Recall also that the first filter at any angle will allow almost all the light through. Since un-polarized light is vibrating in all dimensions, this means that the polarizer does not filter out all the orientations except for those corresponding to the orientation of the filter. It means that the polarizer changes the orientation of almost all the photons to fit whatever orientation the first filter has. So, the first polarizer is not a filter in the sense of blocking some of the light while allowing some to pass through. It is an orienter of the light in a specific plane of vibration.

The State of Confusion in Quantum Mechanics regarding Entanglement

The only way to make sense of Bell inequalities is through the Copenhagen Interpretation – although Bell said that the Aspect experiment confirms the Bohmian Interpretation.
1>In Copenhagen interpretation, each particle has no definite state (spin, polarization, etc.) or the particle has all possible states until measured. When a measurement is made on one particle, the other particle will assume the opposite complementary state instantly (zero time). If one particle is spin up (0), the other will be spin down (180) or if a photon is polarized at 0-degrees, the other is polarized at 90-degrees.
2>However, the Copenhagen interpretation ignores the fact that when particles are entangled initially, they taken on information for complementary states and acquire matching DNA. Bohmian Mechanics assumes that the particles take on a definite state when entangled – Bohm said we are just ignorant of that specific state until we measure it. Otherwise, what does entanglement mean? Twin particles hatched in the quantum foam are said to have complementary characteristics – such as matter vs. antimatter. Therefore, the EPR objection is valid – the particles probably acquired the matching DNA when they were entangled locally – not at the time they were measured.
3>The only way to know if the particles are communicating new information at a distance after they are created, is to change the state of one particle to see if the other follows suit.
4>But then, we are told that to change a particle after entanglement causes decoherence or destroys the entanglement or the change is not communicated for whatever reason. That really means that you get one shot at it. After you change one particle and the other particle follows

suit, then decoherence sets in and the entanglement is lost.

5> The fallacy in this line of thinking is that when measuring a particle in the Copenhagen interpretation, the experimenter is indeed *changing* it from an indeterminate state to a determined state – although the experimenter is not supposed to be able to dictate which state of a number of possibilities the particle will actually take. However, using polarizing filters to measure the orientation of photons can indeed change the orientation of the photon. For example, if a photon is polarized at 0-degrees, it has a 50% chance of repolarizing to 45-degrees if passed through a polarizing filter. This change should be reflected in the other particle's polarization.

6> Aspect tried to rule out locally-determined information by inserting a rotating switch which could rotate faster than light could travel to the filters, so that the photon could be redirected to another detector after the particle was in flight. The purpose of this device was to show that even if the particle was rerouted, it still showed a correlation with its twin particle. However, if the particle was not changed in its polarization, the correlation of the particles could still be attributed to the information embedded in them at the time of entanglement. If Bohm and EPR are right and the particle already had a definite state at entanglement, then the Aspect experiment does not separate local effects from distant effects as advertised. In the Many Worlds interpretation, particles entangled on earth would have their entanglement broken by decoherence when the universe splits. However, Deutsch would probably say that the twin particles remain entangled in different universes in every state possible.

7> The only way to show that the particles can communicate at a distance is to change one particle to see if the other follows suit. Again, unless one of the particles is changed, there is no way to differentiate between local determinism (at the time of entanglement) and distant determinism after entanglement. As one can see, there is much room for interpretation here as to whether Aspect and others have closed any loopholes at all. The loophole-closures depend on which interpretation of quantum mechanics one starts with. If one starts with the Copenhagen Interpretation, then perhaps Aspect has closed a loophole. If one starts with Bohmian Mechanics or EPR, then Aspect has not closed the loophole on local determinism. This example shows how an experiment can lead to a self-fulfilling corroboration because of the theoretical assumption one starts with.

The Confusion over Entanglement leads to the Confusion about Quantum Communication

This confusion about whether "spooky action at a distance" is determined by *changing* the state of a particle or *measuring* the state of a particle has led to the confusion over whether these entanglement effects can be used as a form of communication. Compare these two contradictory statements to witness the confusion:

Templeton: Researchers at Stanford University have taken another major step toward using quantum entanglement for communication, streamlining the process by which two particles can be forced into an entangled state. Once entangled, each should react to changes in the other's quantum spin — if one switches from up-spin to down-spin, the other should hypothetically do the same, instantly and regardless of the distance between them (2012).

Physicist Chad Orzel…entanglement only works if you ask a particle, "what state are you in?" If you force an entangled particle into a particular state, you break the entanglement, and the measurement you make on Earth is completely independent of the measurement at the distant star (2018).

Again, Orzel assumes the Copenhagen Interpretation, but ignores the fact that measuring the particle involves changing the particle. Besides, other physicists say the quantum rule is that you get to change the particles once before decoherence occurs. As Bohr said, one cannot separate the mind from the experiment – the mind of the observer determines the outcome so that cats can be dead and alive until observed. The observation or measurement makes the cat either dead or alive and does indeed change the state of the cat. Thus, in Copenhagen, the observation changes the particle pair from an indefinite state or superposition to one definite state. There is no measuring a particle without changing it regardless of what Orzel says. Whatever the case, if entanglement really exists, it is a very fragile phenomena which is easily broken.

Nutshell: Imaginary Conversation between Physicist and Skeptic

Skeptic: The only way to know if there is non-local communication is to change one of the particles after the particles are separated to see if the entangled partner changes accordingly.
Physicist: But if you force the particle to change, it destroys the entanglement and the particles can no longer talk to each other.
Skeptic: But how else can you distinguish between non-local determinism and local determinism unless you change one particle non-locally to see if the other particle also changes.
Physicist: You *measure* the particle to see what state it is in – you don't *change* it.
Skeptic: But it is well-established in the dominant Copenhagen interpretation that measuring a particle *changes* it from an indefinite state to a definite state. In Copenhagen, the observer determines the state the particle will be in and thus *forces* it to change.
Physicist: You get one change of the particles before entanglement is broken. The instant you change one particle and the other follows suit, decoherence sets in and there is no more communication between particles.
Skeptic: OK, so you can force one particle to change and see a corresponding change in the other one time. My point exactly – this is the only way you can differentiate local determinism from non-local determinism.

Quantum Theory vs. Relativity: Can information travel faster than light?

Relativity indicates that mass and information cannot travel faster than the speed of light; yet quantum theory, particularly the Copenhagen interpretation, indicates that information can be transmitted instantaneously over long distances (i.e., faster than light) between entangled particles. This contradiction between the two big theories of Modern Physics was part of the Einstein-Bohr debates which culminated in the EPR thought experiment, Bell's theorem and Alain Aspect's experiments with entangled particles. Einstein referred to this faster-than-light notion as "spooky action at a distance." Now the consensus in the physics community seems to be that Aspect's experiments indicate that quantum particles can indeed communicate faster than light. This contradiction, of course, requires major patchwork to reinforce the two great pillars of modern physics, and prevent the whole house of cards from collapsing. The patch offered to harmonize the two theories is that the speed limit of light is not actually violated because no information is being transmitted between the entangled particles since they are part of the same system. However, if there is no information transmitted and they are part of the same system, then the information must have been embedded locally when entangled. In this view, it is rather like there is no space or time between them. Again, we see a misuse of language and semantics. If a change in the status of one particle is correlated with a change in status of another particle instantaneously at a great distance, then information has to be transmitted. Otherwise, we are back to the idea that the particles contained this hidden information at inception. Hence the contradiction remains intact between the two great theories. Besides, what else is there to transmit across space besides mass-energy and information? Normally, we think that information has to be carried by some energy such as electromagnetism or some particle such as a photon or graviton. Are quantum theorists saying that quantum information can be transmitted instantaneously without a medium of energy or mass? Logically, either one or both of these theories has to be wrong - despite Herculean efforts to harmonize them.

Not only is there a conflict between Special Relativity and Quantum Mechanics, but also between General Relativity and Quantum Mechanics. Brian Greene (1999) addresses this incompatible marriage between Relativity and Quantum Mechanics:

There is ample evidence that quantum mechanics and general relativity do not provide this deepest level of understanding…where things are very massive and very small— near the central point of black holes or the whole universe at the moment of the big bang,…we require both general relativity and quantum mechanics for proper understanding. The nonsense often takes the form of a prediction that the quantum-mechanical probability for some process is not 20 percent or 73 percent or 91 percent but infinity. What in the world does a probability greater than one mean, let alone one that is infinite? We are forced to conclude that there is something seriously wrong (Kindle Locations 1974-1982).

Summary of critique of Bell Theorem and Aspect Experiments
1~ Entanglement surely embeds some information into twin particles. For example, entanglement opens a channel of communication between inverse twin particles until the entanglement decoheres. Individual, unentangled particles also have innate information.
2~ It is doubtful that single photon or electron pairs can be emitted. Because of their very small size, one or a pair of photons or electrons would hardly be detectable. It is more likely that many individuals or pairs are emitted at once, i.e., a stream.
3~It is highly questionable that one twin photon is hitting a polarizer at the same time as the other twin. Such a precision in the placement of filters equidistant from the light source would require measurements down to the size of Planck length. Perhaps one could argue that even though they didn't hit the screen simultaneously, their effect on each other was simultaneous, but this is not demonstrated in the experimental set-up.
4~No consideration is given to Wheeler's idea that the observation of the experimenter could go backward in time and create the matching results. Time reversal is a regular feature in other quantum concepts; however, entanglement experiments involve a one-way concept of time.
5~The Aspect (and other) experiments contradict the theory that when twin particles are hatched in the quantum foam, they are born with opposite but complementary DNA that is

inherent (locally determined), i.e., opposite charge; matter-antimatter; opposite spin; one travels forward in time, the other backward. This means that there are local hidden variables operating in the quantum foam that make these particles anti-correlated. This also indicates that there are local innate variables operating in the quantum foam, and if these particles were separated, they would already manifest these opposite, complementary properties. Now, I suppose that Bell might argue that if one of the particles were flipped in its polarization, the other would instant flip its orientation to complement it. According to Hawking radiation, if one particle goes into the black hole, it becomes energy, and the other, flying away from the black hole becomes a real particle.

6~The idea that there is no innate, pre-existing information within and between particles seems to contradict Pauli's exclusion principle that electrons with the same quantum properties cannot occupy the same orbitals. The electrons seem to *know* this rule without being entangled and thus manifest anti-correlation of their behavior - unless every particle in the universe is entangled *a la* Bohm. Aspects experiment appears not to support Bohm's idea that all particles are entangled. Apparently, particles have to be entangled *locally* to show the correlations that are demonstrated at a distance.

Entangled particles are also in superposition because, in the Copenhagen interpretation, each particle does not have a definite spin, position-momentum until one particle is measured which causes the other to assume the opposite, complementary status. For example, if one particle is measured to be spin up, the other will convert to spin down so that they will be inversely correlated and have a definite state. In the Many Worlds Interpretation, the particles would go to separate universes once they are separated by the beam splitter. Thus, when each particle communicates its status instantly to the other particle, they would be in separate universes. However, to the rational mind, their inverse correlation is measured on each end of the arm in an experimental apparatus in this universe.

Does Bell-Aspect's Refutation of EPR Confirm that Consciousness Creates Reality as in the Copenhagen Interpretation

Some physicists believe that Bell and Aspect's theoretical and experimental refutation of EPR vindicates the entire Copenhagen interpretation including the principle that consciousness creates physical reality. The original intent of the EPR paper was to create a *reductio ad absurdum* thought experiment of which Einstein was so fond. To wit, if quantum mechanics is right about instantaneous communication between entangled particles, then relativity which posits that nothing can go faster than light is wrong. *And, since relativity is correct about the speed of light as the limit, then quantum mechanics must be wrong or at least incomplete.* The EPR thought experiment began specifically about the Heisenberg Uncertainty Principle which claims that one cannot know both properties of a conjugate pair simultaneously: position/momentum, time/energy, and angular position/angular momentum. Let's imagine a conversation among the founding fathers of quantum theory to make the points/counterpoints clear.

Heisenberg: In a quantum system with entangled particles, one wave function determines the probabilities of each particle so that if you measure the momentum of one particle, it takes on a definite momentum from a number of possible momentums found in the wave. Instantly, no matter how far apart the particles are, the other particle's momentum takes on an equal and opposite value so that the law of conservation of momentum is fulfilled. This conservation of momentum is required because the particles are in the same unified system. And, according the Uncertainty Principle, you cannot know the position of the particles because you have chosen to know the momentum and you can't know both at the same time.

Einstein: But that would mean that the communication between particles occurred faster than light, and that can't be true since nothing can travel faster than light according to relativity. Therefore, there must have been hidden information implanted in the particles at the time they were entangled locally, and they both manifested this already-present information when measured. Otherwise you have spooky action at a distance with no known force or energy to carry the information. Neither gravity nor electromagnetism could carry the information faster than light. (Today, we might say that there is no known fifth force to carry the information.)

Bohr: Mr. Einstein, you're being too logical, the particles are in the same system, so the speed of light limit doesn't apply.

Bell (later): I have a thought experiment on how this can be tested. If I could alter a property of one particle after it had been entangled with the other and separated, we could see if the other particle would still match that property. If the matching beats random expectation, then we have a case for instant communication between particles.

Aspect (later): I have created a device for testing this action at a distance. I have found a way of changing a photon's polarization after it is separated from its entangled partner. If the twin photon still matches its partner's polarization, then quantum theory is corroborated. After numerous tests, I have found that the matching far exceeds chance expectation.

Bohm: All particles in the universe are entangled in a seamless whole. They were entangled when created in the Big Bang and they have inherent information and consciousness.

Bell: I basically agree with Bohm, and I base my theorem on his interpretation. I believe that the particles' ability to communicate with its twin instantly and mirror its state is based on innate programming and information, but the specific information that is communicated regarding changes in the particles' orientation is not built in at the moment of conception. It is acquired after entanglement.

Skeptic: The first objection I would have to the idea that the instantaneous communication of entangled particles corroborates the Copenhagen idea that consciousness creates physical reality is that proving that particles communicate seems to have nothing to say about human consciousness creating this instant communication. I suppose one might say that the observation of this communication is what makes it happen or actualizes it, but other interpretations of quantum mechanics, such as The Many Worlds Theory indicates that consciousness has nothing to do with it.

Furthermore in the Aspect experiments, it appears to me that we are looking at single variables, rather than pairs of variables in the Uncertainty Principle, and we are assuming that if you know that momentum of one particle, you cannot know the position of that particle, and since we know the momentum of the other particle also, we can't know its position either. In the case of Aspect's experiment, I suppose you are saying that since you know the polarization angle of the two particles, you can't know their angular momentum at the same time – but that is an assumption – something that is not really tested or made explicit. So, what is being confirmed here is that if you change the property of one entangled particle, you change the corresponding property of the twin particle simultaneously – and you assume you change the conjugate. So, the essential thing being tested is not the Uncertainty Principle per se, but faster than light communication between entangled particles in the same system. While some have claimed that the corroboration of instant communication between entangled particles validates the whole Copenhagen Interpretation including the idea that *observation and consciousness alone creates physical reality*, I don't think it confirms that principle at all.

The critical thing here is that you change the particle's polarization orientation with a device, not with pure consciousness in some kind of psychokinetic way, and that the twin particle takes on a matching complementary orientation. There is also the assumption that the particle didn't have a definite position (or other property) before you measured it – that is still unknown. Furthermore, the instant effect on the other particle need not have been caused by your consciousness, but by inherent information (DNA if you like) that was built into the particle at the Big Bang or creation. Entangled particles had this capacity to exchange information long before human consciousness arose in the universe, and Wheeler's idea that human consciousness created the universe retroactively is an untestable idea of absurd proportions. Furthermore, Bell based his theorem on Bohmian mechanics rather than the Copenhagen interpretation.

Here is an acid test of the Uncertainty Principle using entanglement. Using Aspect's or other device, detect the position of one particle at the same time you detect the momentum of the other. If entanglement is true, then you should know the position and momentum of both particles simultaneously thus providing an exception to the Uncertainty Principle.

The Seamless Whole and Bohmian Mechanics

David Bohm, following up on the discarded de Broglie Pilot Wave theory, attempted to provide an answer to the EPR paradox by contending that there are non-local variables and that everything in the universe is connected in a seamless whole – thus everything is entangled, if you will. Since everything is interconnected and part of one system, then instant, distant communication among particles is a no-brainer reality. Moreover, Bohm seems to dismiss the point that particles have to be entangled locally (by being brought very close together) in order to be entangled. Thus, the idea that any two particles from opposite ends of the universe (if there is an end) can communicate instantly would involve some mechanism other than entanglement as conceived. Or, perhaps, Bohm would argue that since all particles were enfolded at the Big Bang, then all particles were entangled from the beginning. Bohm's ideas seem to be driven as much by his philosophy and mysticism as by science. It is ironic that Bohm, who was a quintessential mystic influenced by Eastern philosophy, would develop a theory that would remove some of the mysticism from quantum mechanics and restore a rational picture of quantum reality. The pilot wave preserves causality and classical, intuitive interpretation, but at the expense of invoking non-local variables. But then, so do other metaphysical theories such as the Many Worlds idea.

Now there is evidence from fluid dynamics and the study of silicone oil droplets that Bohm and de Broglie may have been right. They may have been correct, at least in part, in that quantum mechanics can be understood in more classical terms of determinism and a particle having a definite location in space and time. Most of the following information for this comparison of fluid mechanics to quantum mechanics is drawn from Natalie Wolchover's 2014 article in *Quanta Magazine*.

To some physicists, the fluid dynamic experiments suggest that quantum objects are as well-defined as droplets, and that they too are guided by pilot waves — in this case, fluid-like undulations in space and time. John Bell (famous for Bell's Inequalities which mathematically show that local hidden variables cannot account for instant communication between entangled particles at a distance) supported pilot-wave theory, having pointed out the flaws in von Neumann's mathematical proof that there could be no hidden variables – even non-local ones. Bell wrote that pilot-wave theory "seems to me so natural and simple, to resolve the wave-particle dilemma in such a clear and ordinary way, that it is a great mystery to me that it was so generally ignored (Wolchover 2014)." The following are some amazing similarities between fluid mechanics and quantum mechanics, particularly the Pilot Wave interpretation.

1~ In de Broglie's pilot-wave concept of the double-slit experiment, each electron passes through just one of the two slits, but is influenced by a pilot wave that splits and travels through both slits. Similarly, the bouncing oil droplet creates waves that, in turn, guide it through one of the slits while the wave spreads out and goes through both slits. These experiments began in 2004, when Yves Couder *et al* at Paris Diderot University discovered that vibrating a silicon oil bath up and down at a particular frequency can induce a droplet to bounce along the surface. Repeated trials show that the overlapping wave fronts of the pilot wave steer the droplets to certain places and never to locations in between — an apparent replication of the interference pattern in the quantum double-slit experiment that Feynman described as "impossible … to explain in any classical way (Wolchover 2014)."

2~Just as electrons are said to be able to tunnel through barriers, oil droplets can also seem to "tunnel" through barriers, orbit each other in stable "bound states," and exhibit properties analogous to quantum spin and electromagnetic attraction.

3~When confined to circular areas called corrals, they form concentric rings analogous to the standing waves generated by electrons in quantum corrals.

4~Oil droplets annihilate with subsurface bubbles, an effect similar to the mutual destruction of matter and antimatter particles predicted by Dirac. However, like matter-antimatter, they do not literally destroy each other but convert back to the substrate from which they came.

5~In each test, the droplet meanders around, forming a chaotic path that, over time, builds up the same statistical distribution in the fluid system as that expected of particles at the quantum scale. But rather than resulting from indefiniteness or a lack of reality, these quantum-like effects are driven, according to the researchers, by "path memory" (similar to Shelldrake's morphic field which Bohm saw as an expression of the Pilot wave and the seamless whole).

6~Quantum statistics are closely approximated even when the droplets are subjected to external forces. In one test, Couder and colleagues placed a magnet at the center of their oil bath and observed a magnetic ferrofluid droplet. Like an electron occupying fixed energy levels around a nucleus, the bouncing droplet adopted a discrete set of stable orbits around the magnet, each characterized by a set energy level and angular momentum. The "quantization" of these properties is very much like Planck's concept of discrete packets of energy.

7~ If space and time behave like a superfluid, or a fluid that experiences no dissipation at all, then path memory could conceivably give rise to the strange quantum phenomenon of entanglement — what Einstein referred to as "spooky action at a distance." As the particles move along, they feel the wave field generated by them in the past and all other particles in the past." In other words, the ubiquity of the pilot wave "provides a mechanism for accounting for these nonlocal correlations (Wolchover 2014)."

As expected, there is much controversy in the physics community regarding the parallels between quantum mechanics and fluid dynamics with some believing that these experiments bridge the gap between the classical world and the quantum world, thus removing the weirdness barrier that separates the two worlds. Others see the similarities as superficial or coincidental and thus rather meaningless. Even Gerard 't Hooft, Nobel Prize-winning particle physicist at Utrecht University in the Netherlands, who believes that quantum theory and the Standard Model are not complete, said: "Personally, I think it has little to do with quantum mechanics." On the other side of the ledger are John Bush of M.I.T. and other fluid dynamicists who have become convinced that there is a classical, fluid explanation of quantum mechanics. Bush said: "I think it's all too much of a coincidence (not to be true)" (Wolchover 2014). Yours truly believes that, because of social gravity which drives groupthink, the physics community generally would not accept this bridge between the classical world and the quantum world no matter what evidence is produced. This does indeed require a paradigm shift from Copenhagen mysticism to a more rational Bohmian view of the quantum domain. I personally believe that there are probably underlying laws or principles that account for this similarity between the classical realm and the quantum realm and that quantum weirdness comes from a lack of knowledge of the unseen world and the inability to separate measurement effects from the natural behavior of subatomic particles. Bohr admitted that one cannot separate measurement data from reality itself even though it is obvious that measurement, no matter when or where it occurs, causes the collapse of the wave. I suppose he would contend that we cannot know this natural behavior without measurement, so we have to accept measurement as reality. Since a sacrosanct principle of physics is to unify all the forces and theories of physics, why not unify quantum physics with classical, Newtonian physics? I suspect that there is a continuity between the micro world and the macro world that is more comprehensible to the rational mind.

> The pilot wave preserves causality and classical, intuitive interpretation, but at the expense of invoking non-local variables - but then, so do other metaphysical theories such as the many-worlds idea. What could be more non-local than a universe splitting and each universe becoming inaccessible to the other.

Does Macro Matter follow Quantum Laws? What does deBroglie say?

Paul Davies (2006) indicates that since macro matter is made of quantum particles, that ultimately it must follow the laws of quantum mechanics. However, in this statement, Davies ignores the holistic principle that at times the whole is greater than the sum of the parts. For example, the water compound is made up of two elements (hydrogen and oxygen) each of which have very different properties from water. Water is inflammable and is used to put out fires. On the other hand, oxygen supports combustion and hydrogen is very combustible. Furthermore, deBroglie showed that even macro matter is associated with waves – so that everything has a wave function. However, there is an inverse relationship between the size of the object and the size of the wave, that is, the smaller the object, the larger the wave; and the larger the object the smaller the wave. The following table shows this relationship.

deBroglie Wave-Particle Concept		
Particle / Object	**Wave**	
	Large	Small
Large		x
Small	x	

Thus, we see that the large object has a very small wave associated with it and would perhaps be less affected by the wave particle duality. However, we might expect that if baseballs were thrown through double slits that they might manifest some of the same wave-particle duality as quantum particles. If a large number of baseballs were thrown or shot through the slits, would they begin to show an interference pattern of dots if their collision with the screen could be marked in some way – say with paint? Perhaps paint balls could be used to show this pattern. If an interference pattern began to build up, perhaps the gap between micro and macro particles would be somewhat bridged.

Zero World's theory and Information theory

Zero World's theory is even more counterintuitive than the other quantum theories, if that is imaginable. Ron Garret (2011), the author of this theory, begins his talk by demonstrating that the Copenhagen interpretation is incorrect because the wave does not actually collapse when measurement is made. He demonstrates this with polarizing filters in which he causes light waves to appear to collapse by rotating two filters perpendicular to each other. Then he interposes a filter at 45-degrees to the vertical one and brings the light wave back. Since the wave does not collapse, he concludes that either the Many Worlds Theory or the Zero Worlds theory is true since neither involve a wave collapse. Since the Many World's theory is so counter-intuitive, he concludes that the Zero Worlds theory is more sensible. How Zero World's is more intuitive and sensible than the Many World's theory is a mystery to me.

Garret (2011) goes on to say that his Zero-Universe interpretation is based on Quantum Mechanics which posits that the classical universe is not "real" and that there is no single classical universe. There is only the quantum universe (which can be viewed as an infinite collection of classical universes). We are not made of hard atoms, we are made of bits of information. The particle-like behavior of quantum systems is an illusion created by the incomplete observation of a quantum entangled systems with a macroscopic number of degrees of freedom. *One classical universe is thus really untenable.* We are our thoughts. We are a simulation running on a quantum computer so that the universe is a good high-quality simulation. What you perceive as physical reality is not actually real; it's actually an illusion (video lecture).

If the physical world is not real, then we have to redefine science because science is based on empiricism, i.e., physical evidence that comes from a real universe. Since Ron Garret background is software engineering, it is not surprising that he sees the world as purely information. To a carpenter, the whole world is a nail; to a computer engineer, the whole world is information made of bits, or perhaps qubits in this case.

Decoherence and Wave Collapse

Decoherence means the process by which quantum information loses coherence and becomes scattered in a sea of particles in the larger world – thus the quantum becomes classical reality. That is, when quantum wave-particles leave the ideal, artificial environment of the laboratory, they mix with other particles in the real world, get scattered, quantum weirdness disappears, and they merge to become the solid reality we experience in the Classical world. The Many Worlds interpretation, unlike the Copenhagen interpretation, posits that the wave doesn't really collapse, but only appears to collapse because of decoherence. At this juncture, we might invoke the holistic principle, that is, the whole is greater than the sum of the parts. Perhaps, we could rephrase that principle to say that the whole (macro matter) is different from its parts (quantum, subatomic matter). There is some justification for this principle in the chemical formula for water. Hydrogen is very combustible, and oxygen supports combustion, but when the two are combined as H_2O, that compound (water) is neither combustible nor does it support combustion. Thus, the whole (water) has different properties from its component elements of hydrogen and oxygen. However, each element imposes constraints on the other to produce this discrepancy between the properties of the parts and the properties of the whole, and if you substitute another element (say sulfur for oxygen), you get a compound with very different properties.

Furthermore decoherence is very much akin to the macro concept of entropy. Entropy means the loss of energy and organization, and decoherence at the quantum level means the loss of information and scattering of particles and waves as they merge into macro reality. However, whereas entropy according to Greene (1999), has the possibility of reversing itself and thus reversing time, decoherence is said to be **irreversible** – that is, is does not permit time reversal. Moreover, delayed choice and the quantum eraser also allegedly permits time reversal. In scientific realism, all of these ideas cannot be correct. How can quantum theory permit time reversal and not permit time reversal at the same time? One has to invoke Eastern mysticism for these contradictions to coexist.

Ron Garret demonstrates how the wave does not actually collapse and can be recovered. By turning a polarizer at a 45-degree angle, the wave can be restored as indicated by the return of the interference pattern. This not only indicates that the wave was not destroyed but that the information associated with the wave was not lost either; thus, no decoherence occurred. If I may expand on Bohm's interpretation, the wave does not collapse, information is not lost and all particles in the universe remain entangled since their initial entanglement at the Big Bang.

Although decoherence and wave collapse are said to be different processes, they seem pretty much the same to me. If I understand the standard interpretation, the wave carries the information and when it collapses, thus revealing the particle, the quantum weirdness is lost, and the particle has a definite place in space and time as in classical physics. The collapse seems similar to decoherence in which waves and particles, interacting with other waves and particles, become scattered and lose their quantum behavior. Accordingly, whether the cat becomes dead or alive in this universe (Copenhagen) or dead and alive but in separate universes (Many Worlds) where they are no longer linked, quantum weirdness has disappeared, and the cat has become classical. Here's what Clegg (2014) has to say about decoherence that sounds so much like collapse:

Decoherence is a one-way affair: once it has washed away quantum-ness, you cannot get it back. The rate of decoherence – the speed at which quantum superpositions vanish increases exponentially as the number of particles in the system increases, so big objects become classical almost instantly. Decoherence, then, makes the quantum-to-classical switch a well-defined process that depends on the precise environmental conditions (p. 52).

> Even though both wave-collapse and decoherence result in the dissipation of superpositions, perhaps the distinction is that in the Copenhagen interpretation, wave collapse is caused by measurement/observation by human beings and in the Many Worlds interpretation, decoherence is a natural process that occurs on its own spontaneously.

> *Everett in his conception of the Multiverse assumed that the wave function doesn't collapse and that the superposition of states lasted forever. However, in the famous cat dilemma, if the universes can no longer interact once a split occurs, the wave must have collapsed or decohered; therefore, the dead and alive cat are no longer in superposition since they are no longer connected by the wave. Each cat has a definite state in two separate universes.*

QBism: Demystifying Quantum Weirdness?

David Bohm, the mystic, tried to demystify quantum strangeness, only to create more mystery with non-local hidden variables and a universe that is totally entangled in a seamless whole. More recently, in 2001, a team of physicists began to develop a model that either eliminates the quantum paradoxes or makes them more appealing to the rational mind. The model, known as Quantum Bayesianism, aka QBism, reimagines the source of quantum weirdness – the wave function (Von Baeyer 2016: p. 94). They key to understanding this weirdness is the mathematical assumption that all possibilities of a particle's location in the wave are equally real and coexist at the same time; therefore, a particle can be everywhere in the wave (which can theoretically spread over the whole universe) until an observation is made. Thus, a cat can be dead-and-alive at the same time if its life depends on quantum superpositions (i.e., the particle has superpositions, not just one position before wave collapse occurs). Problems arise when the mathematical abstraction of the wave function is a real physical entity. The notion that the wave function is not a real physical entity harkens back to Neils Bohr, who considered the wave to be a purely symbolic formalism – a computational tool, merely this and nothing more. QBism, on the other hand, assumes that the wave function is a mathematical blueprint. An observer can use it anticipate how things behave in the quantum world (Von Baeyer 2016).

Quantum Bayesianism is based on the 200-year-old work of English Clergyman, Thomas Bayes. Bayesianism adds an element of intuition and educated guesswork to standard objective probability theory. Objective probability called "frequentism" can be illustrated with the coin toss in which, if the coin is perfectly weighted, the number of heads and tails will be approximately equal in a large number of trials. Frequentism works fine as long as there are fixed probabilities which are repeatable, and, to use another example, as long as the dice are not loaded. However, nature rarely, if ever, gives us evenly weighted coins and unloaded dice, so in predicting nature (such as quantum phenomena), intuition and educated guesses, based on feedback and updated information, may be more effective in predicting natural events. For example, a physician can consult a manual to predict the probability of a person's life expectancy if s/he has a certain type of cancer at a certain stage of development - based on the average duration of life for patients diagnosed with the same malady. That doctor's prediction would be based on frequentism. However, if the doctor used his/her experience, the results of tests, the patient's general state of health and family history, s/he might improve the accuracy of prediction. Unlike the coin toss or the roll of the dice where there are more or less identical, fixed probabilities, each person is different (unlike quarters) which makes prediction a greater challenge. Since biology involves many quantum probabilities, a doctor of Copenhagen persuasion might say that the patient is 70% alive and 30% dead for the next six months - until the patient is examined at which time s/he will be either dead or alive. Such a prediction would be based upon the assumption that each probability is real, rather than the idea that a probability is an abstraction.

Here are some of the ways that QBism, according to Von Baeyer offers a more rational alternative to the Copenhagen, Many Worlds, and other metaphysical quantum interpretations.
1>The observer employs the wave function to ascribe her belief that properties of a quantum system will have particular values, realizing that one's own actions affect the system and change those properties in inherently uncertain ways.
2>One system or one event can have as many different wave functions as there are observers. After observers have communicated with each other and modified their private wave function to account for the newly acquired knowledge, a coherent worldview emerges. In other words, a consensus gives a complete picture of the wave function in the same manner that if the blind men describing an elephant shared their information, a complete image of the elephant could emerge in each mind.
3>Accepting the notion that the wave is an abstraction and not physically real, QBism combines quantum theory to Bayesian statistics which interprets the wave function as a subjective belief and subject to revision by the rules of this model.
4>QBism says there is no paradox. The wave function's apparent collapse is just an observer suddenly and discontinuously revising probability assignments based on new information. Since the wave function is an abstraction, rather than an observation, the collapse is in the mind, not in physical reality.
5>Reinterpreting Schrodinger's cat-box paradox, QBism takes a rational point of view in indicating that no logic or observation can confirm that a cat is both dead and alive at the same

time. This is so because QBism insists that the wave function is a subjective property of the observer, rather than an objective property of the cat in the hat or the cat in the box. The wave function of this system represents a superposition of "alive" and "dead", but a wave function is just a description of the observer's belief, and the fate of the cat is unknown and the math reflects that uncertainty.

Gordon Hazen of Northwestern University takes issue with Von Baeyer's contention that the wave function is merely an artifact of the mind and not a physical reality. Hazen asks: "But what about the famous two-slit experiment in which the wave function of an electron interferes with the portion of itself going through the other slit? Can my belief about which slit the electron went through interfere with itself? Von Baeyer's response is "Hazen is correct that the wave function for a single electron in a two-slit experiment passes through both slits. QBists agree; they differ from other interpreters of quantum mechanics in their insistence that the wave function itself resides only in the agent's mind and that it does not describe the actual path of the electron. It is a calculational device for determining the betting odds the agent should assign for the outcomes of future experiments to detect the electron and is no more substantial than the number on a laundry ticket (Readers Respond to "Quantum Weirdness")."

My response, as before, is that it is not possible to emit one electron or one photon at a time since they are so miniscule. A physical wave is a reality and the arrangement of dots (clumps of particles) on the screen from heavy concentration to light concentration, indicates that interference of a real wave has occurred. However, the Heisenberg table of probabilities describing the wave function is an abstraction and cannot be said to collapse except in the mind. When the particle is found in a particular location, all the probabilities in Heisenberg's table do not collapse, they just do not materialize.

Nutshell: QBism reinforces the view of yours truly that quantum mechanics, despite denial by some physicists, has led to mysticism and anti-scientific theories. QBism, while adding some subjectivity to estimating quantum probabilities, is a quantum leap above irrational interpretations so familiar in the field. Perhaps quantum physics is coming to maturity after a prolonged adolescence of debating zombie cats and the number of angels that can reside on the head of a pin. Here are the key concepts of QBism that clear up many of the absurdities:
1>The wave function is an abstraction, not a physical reality.
2>Therefore, there is no collapse of a wave since the wave is not physically real – there is only a mental collapse so to speak (a collapse in perception when a particle is found).
3>Therefore, theories that assume the wave is physically real, confuse physics and metaphysics, and come up with absurd notions such as dead-and-alive cats, ignoring the fact that Schrodinger meant this thought experiment as a *reductio ad absurdum* of the Copenhagen interpretation. The key error in logic in quantum theory is assuming that a potential event is the same thing as an actual event and that all probabilities are actual realities.
4>Other interpretations, such as Many Worlds and Zero Worlds, also deny that there is a wave collapse but at the expense of even greater absurdities than Copenhagen. Nevertheless, Many Worlds requires decoherence rather than wave collapse which is essentially a distinction without a difference.

CHAPTER 3: MORE WEIRDNESS IN QUANTUM THEORY

Consciousness, Measurement and Wave Collapse

Again, N. David Mermin informs us that the Copenhagen interpretation properly consists of two distinct parts: There is no reality in the absence of observation, and observation creates reality. The question becomes: Is it the actual observation and consciousness that create or alter reality, or is it that *measuring* it alters reality? In the social sciences, it is well-known that the observation of people in an experiment will change their behavior because they are conscious of being observed and therefore behave more in accordance with ideal standards. But are quantum physicists saying, as Bohm asserts, that subatomic particles have consciousness, or that humans have some kind of psychic powers such as psychokinesis by which they can alter or create reality? It is paradoxical that John Wheeler condemned parapsychology as pseudoscience, yet, based on quantum theory, believed that we created the universe with our thoughts going backward in time. This author believes that such extrapolations from invisible particles is one quantum leap too far. Since measurement devices can also collapse waves or at least alter the behavior of a particle, we would have to assume that machines also have consciousness. In much of science fiction, machines manifests consciousness, but I don't believe humans have produced such a machine at this point.

Not only do some quantum physicists believe that human consciousness affects quantum experiments, but that quantum particles themselves have consciousness. David Bohm is perhaps the champion of this animistic concept that everything has consciousness. Perhaps Bohm was confusing information and consciousness. There can certainly be information without consciousness, to wit, the automatic functions of your body are occurring based on information in the autonomic nervous system without your conscious knowledge or direction. Thank heavens, we don't have to think about breathing or our hearts beating.

Quantum tunneling is actually a misnomer in that the particle does not actually tunnel through a barrier that is assumed impassable, but, according to Clegg (2014: p. 80), tunneling is a result of the quantum particle not having a well-defined position in space (uncertainly principle). The barrier may weaken the wave function but does not collapse it or decohere it. Thus the particle may be actualized on the other side of the barrier without the particle having to pass through it. That implies that the particle did not have to move through the intervening space to get to the other side as classical physics would require. Tunneling is certainly no mystery when it comes to transparency and translucency of materials that allow a percentage of photons to pass through. Furthermore, Rutherford shot alpha particles through gold foil and found that occasionally a particle would be deflected by what he thought to be the nucleus of the gold atoms. Clegg (2014: p. 80) also informs us that electron tunneling becomes a problem with the insulating material of transistors which are very thin. Electrons get through the insulation material making it impossible to turn the device off. To remedy this, engineers use another material which is harder to tunnel through. In all these circumstances, the quantum particle seems to go through the material in a quite classical way. Physicists tell us that there is much more space in an atom than matter and that there is mostly space between the nucleus and the orbiting electrons. If this is true, the tunneling would not need mystical quantum processes to enable it.

Delayed Choice and Quantum Eraser – mystical, metaphysical and physical interpretations

Delayed choice means the experimenter can delay the choice as to whether the detector will be off or on after the particle goes through the slit. The quantum eraser means that, after the particle goes through the slit and finds that the detector is on, it can go back through the slit, erase its record as a wave by going backward in time, change its identify from a wave to a particle and return through the slit as a particle.

Here are the basic steps of the experimental set up which is arranged like any other double-slit experiment except:
1) The detector is placed behind the barrier so that it measures the particle-wave after it goes through the slit. The detector can be turned on or off before the emission of a particle or after the particle goes through the slit.
2) An interference pattern does not appear on the screen when detector is on – thus the quantum manifests as a particle.
3) The experimenter chooses whether the post-slit detector is on or off after the particle goes through the slit.
4) The same result occurs. If the detector is on, no interference pattern appears; if detector is off, interference pattern appears.
5) To sum up, the quantum behaves as a particle if detected; it behaves as a wave, if not detected.

So how does the particle decide to go through one slit and manifest as a particle after it goes through the slit and finds the detector on? A mystical interpretation, giving human-like consciousness to the particle, is that the particle goes through the slit and discovers whether the detector is on or off. If detector is on, the particle, goes back through the slit erasing the information from the wave function and, like a quick-change artist, switches into its particle garb and goes back through the slit as a particle. The standard interpretation is that the particle moves backward in time erasing the previously laid down information as a wave and updates the information so that it manifests as a particle. In the view of this author, there is no time reversal certainly not for the experimenter.

Time 1==>	Time 2==>	Time 3==>	Time 4
Photon emitted	Photon goes through slit as waves and sees detector is on	Particle goes backward through slit erasing information as waves	Particle goes forward through slit again as particle not wave

As indicated in the table, there is no backward flow of time, but a definite sequence. There is possibly a backward process, but it occurs in forward time. Rewinding a video tape is a backward process, but it does not reverse time. If time did not flow in one direction, the experimenter would not be able to create the above sequence. Now, one could use the relativity hedge and claim that the time for the particle is different from the time of the experimenter because of the high speeds of the particles.

There are several interpretations of this experiment that I would question.

1) *The idea of emitting one photon by the laser or other light source.* A photon is a hypothetical particle of infinitesimally small size – too small to be seen even with the most powerful microscopes. It is said to be massless, or perhaps it has too little mass to be measured. Thus, I seriously doubt that even the best laser can emit only one photon at a time. A small stream of photons is being emitted with each firing of the light source. Thus, I believe that several photons are emitted, instead of one, and that these photons interfere with each other as manifested on the screen. It is known that air can scatter particles including photons; therefore, unless the experiment is conducted in a vacuum, there would be some scattering effect and therefore collisions of photons causing interference patterns.

2) *Another problem with this interpretation is that one photon goes through both slits thus producing an interference pattern.* This mystical idea of one photon playing with itself after going through both slits comes from the fallacy that the experimenter is able to produce one photon at a time. In all likelihood the laser has emitted a small stream of photons. Individual dots on the screen cannot be assumed to represent only one photon whose dot would be too small to see. Surely the experimenter is seeing a cluster of photons as one dot.

3) *The idea of splitting a photon into two entangled photons is also fallacious I believe.* Such splitting would violate the conservation principle unless each offspring photon is ½ the size of the original photon. I would assume that the split is supposed to produce two photons the same size as the original. If this is the case, then each daughter photon would have to find ½ its mass or energy from some other source or create it thus violating the principle of conservation of mass and energy.

4) *When the experimenter measures a stream of photons whether before, during or after passing through a slit, s/he is causing either decoherence or collapse of the wave at that point.* The experimenter is interfering with the wave interference thus either collapsing it or making it incoherent. It is not the consciousness of the experimenter or the consciousness of a machine, it is what one does to the particle to measure or detect it that affects its behavior. If one hits a particle with a light wave to determine its position, then one is affecting its future position or behavior.

5) *Another finding of the experiment is that if entanglement is used in attempt to fool the particle, nature continues to follow the principle that if a detector is on to determine "which way" or which slit the particle went through, collapse or decoherence occurs for both particles.* For example, if the detector is used to determine which way one entangled particle went, its twin particle will collapse the wave and manifest as a particle on the screen. This connection between entangled particles is not seen between particles that are not entangled. This is difficult to explain in terms of scientific realism and indicates faster than light communication. Subtle is the Lord or subtle is nature *a la* Einstein.

6) *The idea of an experimenter deciding whether a detector will be on or off after the particle goes through the slit is difficult to believe.* If a particle is traveling near the speed of light, how would the experimenter make such a quick decision when the speed of thought is closer to the speed of sound than the speed of light?

John Wheeler's Participatory Anthropic Principle derived from delayed choice: It is from the *delayed choice experiments* that John Wheeler derived his participatory anthropic principle in which humans are said to have created the universe with their thoughts going back in time. Wheeler's version argues that our observations in the present can affect how a photon behaved in the past. Here is the chain of logic (or illogic), known as a thought experiment, which led

Wheeler to this outlandish conclusion (the following information is paraphrased and condensed from Tim Folger's June, 2002 article in Discover Magazine).

1~Wheeler imagines a cosmic two-slit experiment in which an astronomer (observer) is viewing a distant, bright quasar.
2~Between the astronomer and the quasar are two galaxies whose gravity can bend the paths of the photons (particles of light) on their way to earth.
3~When the astronomer looks at one photon coming around one of the galaxies, s/he causes the photon to behave like a particle and appear as a dot on a photographic screen. When s/he looks at a single photon coming around the other galaxy, the same thing occurs.
4~However, when mirrors are set up so as to bring the photons coming around both galaxies onto a photographic screen at the same time, an interference pattern emerges – thus indicating that the photons are behaving as waves.
5~Again we see the familiar meme that, if observed, a photon behaves as a particle; if not observed before it hits the screen, the photon behaves as a wave.
6~Wheeler's conclusion is that the astronomer's consciousness is determining whether the photon is behaving as a particle or a wave. Since it was supposedly a particle and a wave when it left the quasar, the astronomer's consciousness goes back in time and determines whether the photon leaving the quasar will become a particle or a wave on its way to earth. Thus, human consciousness going back in time has created the universe. According to the same logic or illogic, the scientist's observation of the cat, whose fate depends on a quantum event, determines whether the cat was dead or alive in the past before the observation is made.

Another problem with this analysis would become apparent if there were two astronomers looking at photons coming from the quasar. If one is looking at a single particle, s/he is seeing no interference and no wave; if the other astronomer is bringing two particles together to create an interference pattern, then s/he is seeing a wave, but the wave is seen as a pattern of dots on a screen – not as a continuous wave. Now one might argue that for any photon viewed, the fate of that particular photon (whether it becomes a particle or wave) is retroactively determined by the astronomer who is looking at them, so that the astronomer's thought wave is going backward from the future to determine the past and the cause comes after the effect. Or one could invoke the Many Worlds hedge and say that in one universe the photon is a wave and in the other universe the photon is a particle. Choose whatever epicycle fudge factor you like best.

To make a greater-than quantum leap from a particle to the universe is beyond science in the opinion of yours truly. Even if Wheeler is right and the observer's consciousness determines whether an ancient photon will manifest itself as a wave or particle, the observer did not create the photon to begin with – the photon was already created. So, Wheeler could only claim that the mind of the observer determined the form the photon takes – not whether the photon exists or not. Hence to say human consciousness created the universe from "nothing but thought" is an extreme stretch and over generalization drawn originally from a double slit experiment involving tiny particles.

> Nutshell: No matter where one places a detector to determine which slit the particle went through, the detector will collapse the wave. Even if the "which way" of one entangled particle is detected, it will cause the twin particle's wave to collapse and become a definite particle with specific properties – just like its twin. This finding compels a non-local view of forces, but a

> non-local view is not new – both gravity and electromagnetism manifest non-localism in affecting objects not in physical contact with each other, i.e., action at a distance albeit not simultaneous. However, the fact that the wave always collapses to a particle when detected suggests that the detector itself is affecting the wave and causing it to condense to a particle. Thus, the collapse of the wave has nothing to do with human consciousness.

Wave as information or wave as physical reality

The Copenhagen concept of the wave is an abstraction. The wave is seen as a set of mathematical possibilities, not as a physical entity. In other words, the wave is pure information without a physical carrier. Since the wave is pure probability as to where the particle might be located, rather than a certainty, this interpretation is said to be non-deterministic. Not only is the experimenter not able to determine where the particle is, but there is no certain location for the particle until an observation is made. Bohm, on the other hand, sees the wave as a physical entity that carries information. However, making the wave a physical thing is achieved at the expense of allowing hidden variables, and these hidden variables come at the expense of non-locality. Despite the way these quantum interpretations are configured, this author cannot accept a disembodied wave as a carrier of information and probabilities. All information that we are familiar with has to be carried by a physical wave. Electro-magnetic waves carry our electronic information and sound waves carried our vocal communication. The idea of information without a carrier wave is pure mathematical metaphysics.

Technology and QM

Again we hear the mantra that without quantum theory, modern technology would not be possible. But then, we are compelled to ask the question as to which of the four plus interpretations of quantum behavior does our modern technology depend? As observed previously, quite often theory comes after technology and attempts to explain what is happening in the black box that makes the technology tick. Usually there are multiple interpretations of what makes the machine work. The principle of multiple explanations appears to apply here with quantum theory. To wit, do lasers work in accordance with the Copenhagen interpretation where there would be no laser cutting through metal until an observation is made and the wave collapses, or would the laser depend upon multiple universes being formed when there is a probability of getting laser light wave or not? Weinburg (Greene 2014) has said that without quantum theory, we would be in the dark ages and stuck with industrial age technology and that there would be no lasers or technology that depends on subatomic particles. Surely Weinburg should evaluate that statement in light of historical fact. How many technologies using quantum particles such as photons and electrons existed prior to the development of quantum theory in the 1920s? Let me count a few of these technologies: 1> Photography 1839, Movies 1890s, 2> Telegraph 1830s, 3> Telephone 1876, 4> Radio 1880, 5> Fax 1843, 6> Television 1907, 7> Electric lighting 1882, 8> Early computers using electricity 1880s. And this list does not include the invention of electric generators, electric motors and the many other electrical inventions by Michael Faraday. Furthermore, the Balmer formula was created in 1885 and the Periodic Table was developed in 1869.

Far from living in the dark ages as Weinburg suggests, even without quantum theory, we would have electric lighting, radio, TV, telephones and a host of other inventions that involve quantum phenomena such as the interaction of electrons and photons. Quantum theories, like the planetary picture of the atom, may help us understand what makes quantum technology tick, but the technology does not altogether depend on this picture. Here again, we witness irrational exuberance on the part of scientists eager to promote a theory's importance to technology by making such exaggerated claims. Again, I ask the question as to how any advanced technology such as computers, atomic clocks, and lasers could operate so precisely if subatomic particles behave so randomly, unpredictably and with such great uncertainty. Rather than making technology work or explaining how technology works, these theories of chaos and uncertainty would seem to contradict the precise workings of our modern machines and cesium clocks. Moreover, if these theories are so solid and well-established, then why are there so many widely-divergent interpretations about the behavior of subatomic particles?

CHAPTER 4: QUANTIZING THE UNQUANTIZABLE

String Theory – the Ultimate Quantum Theory

String theory is the failed search for the DNA of the universe and the unification of all known forces.

In the two and a half decades since it first captivated physicists, despite thousands of published papers and the expenditure of billions of dollars, there is no proof whatsoever that string theory is correct. Not one prediction of the theory has been experimentally testable." —Boston Globe quoted by Woit (2006: preface).

Mention has already been made of String Theory and its attempts to unify the known forces of nature with a single theory and a single equation. As indicated String Theory attempts to achieve this grand unification by positing that the elemental reality is composed of one-dimensional strings or filaments that vibrate at various frequencies in 11 dimensions to produce the basic particles that, in turn, make up macro reality and are identified in the Standard Model. The theory is admittedly mathematical, rather than physical, because it is founded on purely mathematical assumptions of one-and two-dimensional matter and multiple dimensions of space, neither of which have ever been observed and for which there is not even indirect evidence. An assumption of one-dimensional matter is surely a false assumption. All matter, so far as observed, is three-dimensional – no matter how fine you cut it. It defies logic to believe that any amount of one-dimensional matter, if there were such a thing, could add up to three-dimensional matter. Now we are told that some beautiful mathematics have come out of this theory, but, if so, it is pure, other-worldly math that has nothing to do with this empirical world. It is what I have called "chess math" – an artifact of the human mind. We are also told that if physicists do not find super-symmetric particles in the next few years, then the theory must be discarded for lack of evidence. I would argue that if physicists do not find one dimensional strings and extra dimensions beyond the three that we experience, then the theory is of little use except possibly in the field of pure mathematical metaphysics.

My critique of String Theory will carry little weight, but when a critique comes from the physics community, then perhaps other physicists will take note. Here is what several physicists have to say about the sociology and the economics that drive the continuing research on a theory that holds little promise. They argue that there is much conformity, group think, and lack of critical thinking about String Theory.

Peter Woit views the state of string theory research as unhealthy and detrimental to the future of fundamental physics. He argues that the extreme popularity of string theory among theoretical physicists is partly a consequence of the financial structure of academia and the fierce competition for scarce resources (2006).

In his book *The Road to Reality*, Roger Penrose (2004) expresses similar views: "The often frantic competitiveness that this ease of communication engenders leads to 'bandwagon' effects, where researchers fear to be left behind if they do not join in (p.1018)." Penrose also claims that the technical difficulty of modern physics forces young scientists to rely on the preferences of established researchers, rather than forging new paths of their own (pp. 1019–1020).

And, Lee Smolin (2006) informs us that "String theory now has such a dominant position in the academy that it is practically career suicide for young theoretical physicists not to join the field. (p xx, Introduction)." (Furthermore) "Some young string theorists have told me that they feel constrained to work on string theory whether or not they believe in it, because it is perceived as the ticket to a professorship at a university. And they are right."

Again, I would argue that the problem with string theory is much more fundamental. Its chief problem is the domination of pure mathematics in physics and the faulty use of language and logic to describe nature. Consider what String Theorist Michio Kaku (2013 says about this wedding of pure, other-worldly math and physics.

It turns out that 100 years ago math and physics parted ways. In fact, when Einstein proposed Special Relativity in 1905, that was also around the time of the birth of topology, the topology of hyper-dimensional objects, spheres in 10, 11, 12, 26, whatever dimension you want, so physics and mathematics parted ways. Math went into hyperspace and mathematicians said to themselves, "Aha, finally we have found an area of mathematics that has no physical application whatsoever." Mathematicians pride themselves on being useless…and they said the most useless thing of all is a theory of differential topology and higher dimensions.

…recently we discovered string theory, and string theory exists in 10 and 11-dimensional hyperspace. Not only that, but these dimensions are super. They're super symmetric. A new kind of numbers that mathematicians never talked about evolved within string theory. That's how we call it "super string theory." Well, the mathematicians were floored. They were shocked because all of a sudden out of physics came new mathematics, super numbers, super topology, super differential geometry (Kaku 2013).

It's interesting that Kaku thinks that this is a good development – that physics adopted concepts (hyper dimensions) from math that have no known physical reality and made them a part of physics which is supposed to describe physical reality. This admission by Kaku sums up the problem with String Theory and modern physics in general – pure, other-worldly math has been adopted to explain this world. Where, in the physical world, does one find 1-D strings and 7 extra dimensions of space – these are pure mathematical constructs borrowed from the mathematics of topology. This development in physics is truly imagination without a strait jacket. If math and physics are the same thing, then why do we need physics?

To sum it up, the problem I see with String Theory is that it is pure, metaphysical mathematics and probably has nothing to do with the empirical world. It is interesting that so many physicists would invest their whole careers in a theory that has so little chance of being tested empirically. Here is a summary of the problems with the theory.

1) *To begin with, there appears to me to be a logical error right up front.* A string in the form of a loop, either closed or open, would be two-dimensional, not one-dimensional. One dimension would be a straight line – any curved line manifests two dimensions. Of course, any piece of matter, no matter how small has to be three-dimensional. No amount of 2-D matter will add up to 3-D matter and that is the only kind of matter ever observed. The idea of 1-D and 2-D are pure mathematical and geometric concepts – not physical reality. In other words, they belong in the metaphysical domain of the mind, not in objective, physical reality.

2) *String Theory involves **massless particles** such as photons, tachyons and gravitons.* The concept of a massless particle is a direct contradiction of Einstein's sacrosanct principle $E=MC^2$ which indicates that energy and mass are interchangeable, and that energy has mass. Even if particles are divided into energy/force particles and matter particles, then both have mass. The assumption of a no-mass particle is, I suppose, to make the math work – a case of the tail wagging the dog. In the case of the tachyon, it is mathematically necessary for this particle to be massless, because since it travels faster than light, it would violate Einstein's theory that nothing (no mass) can go faster than light. Thus, physicists can have their cake and eat it too. They can have a particle that exceeds the speed limit of light because it is massless. Likewise, since a photon travels at the speed of light, it must be massless also.

3) *String Theory is not testable, non-falsifiable and therefore it is not a scientific theory.* The theory deals with tiny strings and dimensions that are billions of times smaller than the atom. Since scientists have never seen an atom with the most powerful electron microscopes, it is unlikely that they could ever observe or test anything so small. One proposal to test the theory is to smash particles together in a particle collider to see if energy is lost. If energy is lost from the collision, then it would be inferred that the energy went into other dimensions. This might be evidence for other dimensions, but certainly not the only explanation for the loss of energy. And, which of the six extra dimensions would the energy have gone into? If it went into a fifth dimension, that would not prove the existence of the other six hyper dimensions. Surely, when the physicist is dealing with such a small loss of energy, there would be a measurement problem which could always be questioned for accuracy. Would the energy have escaped into the fifth dimension or would it have escaped measurement.

4) *Another complication of string theory is the addition of Supersymmetry which creates Superstring theory.* Supersymmetry requires **supersymmetric particles** known as superpartners or sparticles. According to the supersymmetry theory, each fermion should have a super partner boson which would be a mirror image of the fermion, and each boson should have a super partner fermion (Quigg 2008). To date, no such super partners have been found at CERN or the Fermi Lab. Furthermore, Michio Kaku, a string theorist, says that it would take a particle collider the size of the Milky Way Galaxy to test String Theory – in other words, it is untestable.

4) *The theory is essentially **numerology**.* It is based on shaky assumptions at best: hypo-dimensions such as 1-D strings and 2-D space and up to six hyper-dimensions combined with time. The mathematics based upon these imaginary dimensions can be no better than the assumptions upon which they rest. No complex, differential calculus can ever make these assumptions right. Two physicists worked on the math for years to overcome the anomalies (or contradictions) that came out of different equations (Greene 2016). Without testability in the empirical world, how can anyone be sure that the numbers were not manipulated in such a way as to yield the right answer? Again, this is pure math and probably does not describe anything in the physical world. A mathematical proof is not the same thing as a physical proof. This Herculean

exercise in mathematics is reminiscent of Einstein's finagling of mathematics in his many attempts to calculate the perihelion shift of Mercury with General Relativity without contradicting Special Relativity.

5) *String Theory appears to me to use one conjecture to explain the anomaly in another conjecture.* Thus, speculation gets piled on speculation and the strait jacket that disciplines imagination is lost. For example, from sub-microscopic strings and dimension come branes (from mem<u>branes</u>) where parallel universes may exist in different dimensions or different slices of spacetime. I can't claim to know the mathematics that gets you from tiny curled up dimensions to huge dimensions large enough to contain a universe; however, if a whole universe exists in another dimension or brane, is there only one dimension in that universe? If so, now we are back to one-dimensional matter, and no matter, other than the three-dimensional kind has ever been observed. This is indeed speculation piled upon speculation.

6) *The evolution of String theory to Supersymmetry to Superstring Theory to M-theory to branes and multiverse is a set of fixes for the problems in the earlier theories.* Here is what Feynman had to say about these epicycle-type fixes.

I don't like that for anything that disagrees with an experiment, they cook up an explanation—a fix-up to say, "Well, it still might be true." For example, the theory requires ten dimensions (Woit 2006: p. 174).

One rationalization for the lack of success in String Theory is that the mathematics is said to be too complex even for the world's most brilliant mathematicians to calculate. The attempts to describe the single vacuum state of the universe has resulted in 10^{500} possible equations. Rather than seeing this as a weakness since there can be only one vacuum state for this universe, Susskind applies an epicycle-hedge to say these 10^{500} possible states, rather than describing one universe, really describes all the possible universes that could exist with different constants and natural laws. Hence, Superstring theory mathematics tell us that there could be 10^{500} universes out there that are part of the multiverse. So, a failure begets a new success with the help of Ptolemaic epicycles. Peter Woit (2006) points out some more numerologies that can't be reconciled with experiment.

The...MSSM (minimal supersymmetric standard model) has at least 105 extra undetermined parameters that were not in the standard model. Instead of helping to understand some of the 18 experimentally known but theoretically unexplained numbers of the standard model, the use of supersymmetry has added in 105 more. As a result, the MSSM is virtually incapable of making any predictions. In principle, the 105 extra numbers could take on any values whatsoever, and in particular, there is no way to predict what the masses of any of the unobserved superpartners will be (p. 168).

It is interesting that so many physicists have taken String Theory to task at some risk to their careers. Many of these physicists have used very technical language to refute String Theory on its own territory - as we see in this excerpt from Peter Woit's book (2006).

An especially dangerous problem for the theory is something called the doublet–triplet splitting problem. The supersymmetric grand unified theory must contain not just the usual Higgs doublets, which are in the two-dimensional representation of the standard model SU(2), but also Higgs triplets, which are in the three-dimensional representation of the standard model SU(3). That is, they come in three colors, just like quarks. The mass of the Higgs doublet must be small, and the problems in the MSSM that come from trying to arrange this have been discussed earlier (p. 170).

Wouldn't it be much simpler just to say that String theory is based on the assumption that there are one-dimensional strings that vibrate in 11 dimensions – none of which has ever been observed and has a very slim chance of ever being observed. All the rest of that technical stuff rests upon these root assumptions. If these assumptions are untrue or untenable, then all the convoluted language and complex math is irrelevant – and much ado about nothing.

To sum it up, String Theory is essentially pure math, borrowed from the topology of higher dimensions. It is not physics or science and has little likelihood of ever becoming physics. Much money and energy has been spent on a theory that is not scientific. This problem in string theory is symptomatic of a larger problem in Modern Physics of the trend away from physics into metaphysics, from empiricism to pure rationalism. This trend is more apparent in string theory because it has carried this thought process to its illogical extreme.

Quantizing space, time, spacetime and gravity (Loop Quantum Gravity)

In attempting to unite the forces of nature and reconcile quantum theory with relativity, modern physicists believe that it is necessary to quantize space and time in order to quantize gravity. Since rejecting the ether as a medium to carry the electro-magnetic force and perhaps to carry gravity, physicists have been on a quest to find another medium to carry gravity in order to avoid the theoretically-troublesome "action at a distance". Einstein attempted to resolve the "action at a distance" conundrum by imagining space and time or spacetime as a medium that can be warped or curved in order to account for the orbiting behavior of smaller bodies around larger bodies. Thus, by treating spacetime as a material thing, rather than an abstraction, Einstein believed that he had found the medium that connects the planets to the sun in a physical way, causing the planets to follow their star like sheep following a shepherd. Thus, Einstein could reject quantum entanglement as "spooky action at a distance" without appearing contradictory. However, to make space and time or spacetime as a physical medium presents philosophical and semantic problems. If space is vacuum as Einstein indicated in Special Relativity and time is a metaphysical concept, then spacetime cannot be conceived as a physical medium. To say that the warpage of a metaphysical construct (spacetime) causes the planets to orbit around the sun is tantamount to saying that the longitudinal lines on a globe cause the earth to spin. Apparently, some physicists are dissatisfied with making a metaphysical concept into a physical medium and are searching for a new ether that would physically connect planets to a star. Loop Quantum Gravity is an attempt to quantize spacetime, by finding atoms or quanta of spacetime in the form of loops that link to each other and link orbiting bodies to their orbited bodies. Here are my thoughts on quantizing spacetime and gravity.

If space is vacuum or nothingness as Special Relativity indicates, then the idea of finding a discrete unit (like a Plank constant) of nothingness reduces to absurdity (space can be measured by feet or meters or any other standard arbitrary units). If time is a metaphysical construct (a thing of the mind), then any attempt to find a natural, discrete unit of time is meaningless (the second can be defined in any number of ways). Likewise, to find a discrete unit (like an atom) of spacetime (an abstraction) is even more problematic. If both space and time are non-physical constructs, it is illogical to think that the spacetime combination takes on physicality.

Near the end of writing this book, I happened upon a new insight about space. In a sense, space is almost filled with quantized particles, namely photons. Consider the fact that in our orbit around the sun, we always see the sun's light no matter at what point in our orbit or at what point in the sun's orbit around the galaxy we view the sun. Not only is this true of the sun, but it is also true of the distant stars. No matter where the earth carries us in space, we can see the North Star in the Northern Hemisphere. The omnipresence of light (photons) is made possible by the fact that stars are spherical and thus emit light in every direction (360-degrees). Consider also the Cosmic Microwave Background Radiation (CMB) that is present everywhere in space. The attractive part of this conjecture that photons may carry the gravitational force and the electro-magnetic force is that it would unite electro-magnetism and gravity in that the same particle carries both forces.

After writing this passage in the book, I discovered that Andrei Sakharov had a similar insight and was endorsed by John Brandenburg who also believed that photons carry the gravitational force and thus unite gravity with electro-magnetism. Perhaps if there were low density photons in one area of space and high-density photons in another, there would be a pressure to drive orbiting bodies in the manner that high atmospheric pressure drives winds which move massive objects. Varying photon density could possibly account for the Casimir Effect that drives uncharged plates together in a vacuum. The plates are supposedly driven together by virtual particles coming out of the vacuum or zero-point energy (vacuum energy) of a quantized field (Jaffe 2005). Since photons are omnipresent, there is no need to invoke imaginary virtual particles coming from nothing since actual photons can inhabit what is considered a vacuum. Particles other than photons that appear to be emerging out of nothing might be from the CMB and radiation given off by stars, particularly super novae. Thus, ever-present photons rescues physics from the fallacious notions of something coming from nothing and zero point energy.

In this late conjecture about photons as potential carriers of other forces, the question becomes whether photons are strong enough to carry not only the electro-magnetic force but the gravitational force as well. Photons can be a powerful force in carrying the electro-magnetic force which is said to be stronger than gravity. Furthermore, photons can drive a space vehicle that uses a sail to catch the rays of the sun, and photons can cut diamonds and hard metals if they are concentrated as in lasers. However, photons used in these ways are repulsive, rather than attractive forces as in when a sail might be used to power a spacecraft. Thus, the sun would repel the planets rather than attract them if the sail model were used to explain gravity. Likewise, the CMBR is said to be just above absolute zero, and heat is a repulsive force as well. Unless light can be induced to be a standing wave, it is unlikely that it is a candidate to carry the gravitational force – even though it is the only known particle that is omnipresent. However, on the other side of the ledger, the Casimir effect has been found to be both attractive and repulsive (Munday *et al*

2009). If the cause of the Casimir effect is actually real photons inhabiting space, rather than imaginary virtual particles, then differing densities of photons could account for attraction as well as repulsion.

Photons are said to be massless; therefore, if this is true, they would not qualify as a physical medium. However, I have countered this notion of massless particles by invoking Einstein's famous formula ($E=MC^2$) which indicates that energy and mass are interchangeable. Hence any particle that carries force or energy has to have mass as well. Logically, there can be no force particles that are massless. Even though it appears that photons are everywhere in space, there is still a distinction between space and the thing that is occupying space. The earth occupies space, but the earth itself is not space. Similarly, an electron occupies space, but it cannot occupy just any space in its orbit around the nucleus of the atom. There are spaces it cannot occupy and so electrons take a quantum leap from one shell to another.

Furthermore, since photons or light waves are omnipresent in space, it seems logical that some would be of the same frequency and would overlap each other and create patterns of constructive and destructive interference similar to those seen in the double slit experiment. If this is so, then these shifting patterns could have implications for the red shift effect which is the foundation of the Big Bang theory. Furthermore, rather than looking for gravity waves in the virtual vacuum of space, physicists should observe the tides which are created by the sun and moon's gravity. If gravity requires a medium to express its wave properties, then water is a fine medium for that expression here on earth.

Since photons are essentially omnipresent in space, perhaps they partially account for the apparent quantum foam which gives rise to such ideas as vacuum energy, virtual particles, borrowing energy from the vacuum. No particles have to be created from the vacuum thus violating the conservation of mass and energy laws, the particles are already there.

Light as a waving stream or streaming wave

In the view of this author, physicists have chosen the wrong metaphor for describing the wave-particle duality of light and other particles. In an ocean wave or sound wave, the image is that of energy creating a disturbance in a passive, still medium (water or air). Light or electro-magnetism is different because it does not need a pre-existing medium (such as ether) to be propagated. Since light is particulate, it can move through a vacuum because the particles can carry the waves without assistance from any other medium. Hence, light is more like a river or stream where moving molecules (as opposed to still molecules) of water carry a wave. The less external medium (such as atmosphere) there is, the better light moves and light achieves its maximum speed in the absence of a medium (i.e., in a vacuum). Actually, water waves and sound waves can be projected into a vacuum and demonstrate waves. If one shoots a stream of water into a vacuum and varies the pressure of the stream, water waves will occur, and if a blast of air is shot through what was a vacuum, it, too, can carry sound. Of course, whether one shoots light, water, or air into a vacuum, it is no longer a vacuum, but the point is that all these media can show wave properties when travelling through a near vacuum and there is no essential difference between these phenomena in wave properties related to the vacuum when comparing apples to apples.

This view of waves and particles is that there is no duality and no uncertainty. Just as water molecules and atmospheric particles (or other solid media) carry sound waves so do quantum particles carry electro-magnetic waves. The double slit experiment with its mystical interpretations reveals that waves and particles are never separated in their manifestations. Whether one shoots particles through a single slit and gets a scatter gram of dots or shoots particles through a double slit and gets a pattern of dots suggesting waves, in both instances, you still get dots on the screen. Only when you run the particles through the double slits do you get the separation of the particles into lines and spaces which manifests as interference patterns on the screen. Now, some, resorting to mysticism again, want to say that the particles convert to waves when going through the double slits and then convert back to particles when they hit the screen. This is a totally unnecessary interpretation of the facts since the pattern of particles is wavelike indicating that the particles were carrying waves as in a waving stream. The ether theory of a static field like a static lake is the wrong metaphor, particles in motion can also carry waves, not just still particles in a pacific ocean.

A good example of particle-wave unity is a flow of electrons in a wire or perhaps shooting electrons in a vacuum tube as J.J. Thomson did. The stream of electrons going through the cathode tube (which is a near vacuum) shows that the electrons do not need an ether-like medium to travel through – unless one wants to invoke the quantum foam as a medium, but then a flow of electricity would be a rhythmic disturbance in this field like a water wave – there would be no need for separate particles besides those in the pre-existing quantum foam. Of course, such a contention would contradict Thomson's findings that electrons have mass; thus they are not just a wave disturbance in the quantum foam. In addition, it seems obvious that when electrons are moving through a wire, they are creating a field of waves which can be detected and measured. There we see the particle and the wave manifesting themselves simultaneously while theory says that the wave and particle always manifests themselves separately – never together. The electron with its wave intact again demonstrates that electro-magnetism is a moving stream with waves, creating its own medium as it goes.

However, Cox and Forshaw (2014) are at odds with my interpretation as they take the conventional Copenhagen interpretation. Here's what they say:

This is an electron wave, and not a wave of electrons: we must never fall into the trap of thinking otherwise...But we do need to explain why the electron pattern is made up of tiny dots as the electrons hit the screen one by one. At first sight that seems in conflict with the idea of a smooth wave, but it is not... Put another way, when we said that the electron is 'somewhere within the wave' we really meant to say that it is simultaneously everywhere in the wave! (Kindle Locations 460-487).

To put the critique of Cox and Forshaw's arguments in a nutshell:
1. **They say that electron wave behavior is due to an electron wave, not a wave of a bunch of electron particles zigging and zagging rhythmically as they travel.** Quantum physicists

> say that electron is a tiny magnet which has a magnetic field involving north-south magnetic poles. If this is so, perhaps there is an electron wave. More on this later. However, a stream of electrons, each with a magnetic field, probably wave together in resonance, like a river that involves a flow and a wave of molecules of water. Since photons and electrons do not need a medium to wave in, they probably create their own medium in which to wave.
>
> **2. Cox and Forshaw compare an electron wave to water wave of the type that involves creating a wave in a still pool or pond.** Of course, that analogy breaks down because a wave through water needs a pre-existing medium (water), and particle waves can exist in a vacuum (without pre-existing medium). However, some physicists might say that the vacuum is not really a vacuum and there is a quantum foam which might be considered a new ether in which waves of many frequencies can be generated. However, Cathode rays (stream of electrons), as indicated above, can travel through this quantum foam if it exists, manifesting particle and wave properties simultaneously. Unlike water waves which need no other particle, electron waves would involve a particle traveling through a quantum foam of particles and energies.
>
> **3. Cox and Forshaw indicate that the dots on the screen in a double-slit experiment do not indicate that electrons are particles.** This really goes against rationality and commonsense, and quantum physicists seem to prefer a non-rational interpretation. Again J.J. Thompson's experiment indicated that streaming electrons are particles that have mass which can be measured – thus electrons would make dots on the screen since they are particles with mass. Furthermore, that stream of electrons involves a field which reacts to a magnet. What we actually see on the screen are *dots*, but the patterned distribution of dots looks like a wave with areas of high concentration of dots (crests or constructive interference) and areas of low concentration of dots (troughs or negative interference).
>
> **4. The electron is not one place in the wave, it is everywhere simultaneously in the wave.** They contend that it is not just that the electron has a possibility of being in all places at once, but the electron actually is in all those places at once and that is what causes the interference pattern - even when a single electron is sent through the double slits. Of course, I believe it is impossible to emit only one electron or photon at a time. Furthermore, they state that an electron wave, unlike a water wave, does not involve a physical disturbance. Instead of a physical disturbance in a medium, an electron wave, like other quantum waves, is a statistical set of probabilities of where the electron might be. Of course, a set of probabilities is pure mathematical abstraction, not a physical entity. Furthermore, if the electron is everywhere in the wave at once, then there is no need for a probability calculation to describe where it might be – it is in all locations. So, if someone asks where the electron is in the wave, then the answer would be that it has a 100% chance of being everywhere in the wave at once. Thus, there is no probability, the location of the electron is *certain (it's everywhere in the wave)*, not probably one place or another, and then there is no uncertainty principle. Of course, this interpretation may make too much rational sense in a field of physics that seems to prefer the counterintuitive and mystical over rationality.

This idea of an electron being in all possible places at once is also expressed as an amorphous electron cloud where electrons exists as potentials, not real particles. However, if this were true, then it destroys the whole quantum theory of electron shells or orbits. These orbits which are portrayed as rather definite places that admit only a certain number of electrons with differing quantum numbers provide the explanation for electro-magnetic radiation. Physicists tell us that when an electron in a higher orbit jumps to a lower orbit, a photon is emitted, and conversely when an electron jumps from a lower orbit (or energy level) to a higher one, a photon is absorbed. It is difficult to reconcile this picture of the atom where electrons have a rather definite place in concentric orbits with the picture of the electrons in some amorphous cloud surrounding the nucleus with no definite location. Now, those of the Copenhagen persuasion might argue that the electrons do not have a definite place or state until they are measured or observed, and then they take on this well-organized particulate form. Of course, this contention would beg the question as to how electrons and atoms functioned in this well-defined state before humans evolved thus creating the chemistry and physics that makes life possible. But then, Wheeler says that our observations in the present can be retrograde and actualized things in the past before our human appearance on the earth. As we see, there is no way to falsify a theory that is insulated from scientific empiricism.

I would further argue that in the quantum field, the wave and particle *appear* to manifest themselves as one form or the other, but never as a wave and particle at the same time, not because that it is the innate nature of the quantum, but because of measurement. If a particle is hit with light to determine its location, then the properties and trajectory of the particle have been changed. Furthermore, if the particle is everywhere embedded in the wave, as Cox and Forshaw indicate, then the particle and wave coexist simultaneously. Only when one tries to measure the particle in an obtrusive way that directly affects it, does it develop multiple personality disorder.

The Interaction of The Quantum World and The Macro World

A la quantum mechanics: Every living creature is in a quantum state of dead-and-alive, not just the ill-fated cat

Mention has already been made of the fact that many physicists take Schrodinger's cat box thought experiment literally - even though Schrodinger meant it as a parody and *reductio ad absurdum* of the Copenhagen interpretation. Now this oxymoron of a cat being dead and alive no doubt comes from the notion that a subatomic particle exists as both a particle and a wave until an observation is made – at which point the wave collapses and, as a quick-change artist, reveals itself as a particle. Apparently, no thought is given to the rational notion that when you measure something physically by interfering with it, you change the behavior of the thing measured. Thus, it is not your consciousness which collapses the wave; it is your physical interference with the wave. Whatever the case, one would think that the evidence for a dead-and-alive cat must be overwhelmingly compelling for rational people to willingly suspend disbelief and accept such an outlandish prediction. Let's imagine a debate between a rational-empiricist and a quantum mystic and see how the arguments might shake out.

Rational-Empiricist: This hypothesis of a dead-and-alive cat violates the scientific method and the philosophy of science which is based on rational-empiricism. It violates rationalism in the fact that contradictory statements cannot both be true. The idea of a dead cat that is also alive is a self-contradiction. This non-contradicting principle in the scientific method requires that hypotheses be falsifiable. The hypothesis is set up as contradictory statements, and the evidence must support one or the other prediction, not both. Thus, a rational hypothesis would be that the cat is either dead or alive, but not both dead and alive. The dead-and-alive prediction is unfalsifiable because the original Copenhagen prediction was that the observation makes the cat, which was dead-and-alive, either dead or alive when the box is opened, and the cat is observed. Therefore, the observation, in this experimental design, would support the prediction of dead-and-alive no matter which observation was made. If the cat is observed to be dead, then Copenhagen would say the cat was dead-and-alive before the observation, and if the cat is observed to be alive, then Copenhagen would say the cat was dead-and-alive before the observation. Thus, either observation (cat dead or alive) would be interpreted as support for the prediction, even though the state of being dead-and-alive could never be observed.

Quantum Mystic: Well, one might argue that life-and-death is not a duality but that it is a continuum, so that the cat could be in an indeterminate state hovering between life and death – and whether the feline is dead or alive (or something in-between), depends upon how one defines death.

Rational-Empiricist: But, if the cat is in a zombie state between life and death, it would mean that the release of poisonous gas has already been triggered by a quantum event, and the cat is in the process of dying. Falsification depends ultimately on an *either-or* duality, not a *both-and* unity. Also, the Copenhagen interpretation is that the cat is in a fully-alive, healthy state and fully-dead state before the observation is made.

Quantum Mystic: But many sound theories in physics predict things that cannot be observed directly. Take the inference that there are planets orbiting other suns. Even though these planets may not be observable, one might observe a wobble in the spin of the star indicating a gravitational tug, or one might see a shadow on a star presumably cast by a planet. So, astrophysicists reasonably predict that these planets exist without seeing them.

Rational-Empiricist: Yes, that is a reasonable prediction based on sound evidence and the observation that our own sun has many planets, so there is a small leap of inference (or leap of faith, if you like) to conclude that planets exist around other stars. However, this reasonable inference, based on high probability, does not indicate that the planet both exists and doesn't exist at the same time until it is observed. Such a claim would be mysticism. Moreover, every living organism, not just the cat in an experiment, is in a quantum state of uncertainty. This dead-and-alive state must be universal according to this theory because every living being exists in a world of widespread radiation that is potentially deadly and can cause mutations resulting in ultimate death. Moreover, since we are made of quantum particles, our whole being must reside in a quantum state of uncertainty. Hence, we could say that we humans, like the poor cat and all other creatures, exist in a quantum dilemma of being both dead-and-alive at the same time. Perhaps the "Wanted" posters in the Old West should have read "Wanted Dead and Alive", rather than "Wanted Dead or Alive".

Quantum Mystic: But in the Many World's theory, the cat is allowed to be dead and alive at the same time because the wave doesn't collapse, but, instead, the universe splits so that the cat is dead in one universe and alive in another. So if the cat is observed to be dead in the experiment, it is alive in another universe, and, vice versa, if the cat is observed to be alive in the experiment, it is dead in the parallel universe.

Rational-Empiricist: Even with this fix, you have the same problem of unfalsifiability which is crucial to the scientific method. As you said, in whichever state you find the cat, the cat is in the opposite state in the other universe. So, if the cat is alive in your experiment, you say that your prediction was right, and if the cat is dead, again you say you were right – either outcome supports the same prediction. Furthermore, you say that the other universe cannot be observed because of decoherence, so again, you have immunized your hypothesis against falsification. And, wouldn't we see some diminution of matter in our universe (even as large as it is) if the universe splits countless times every second due to quantum probability waves? On the other hand, if you say the universe makes a copy of itself to accommodate every probability, then you are violating the law of conservation of mass and energy.

Proposed Experiment: Not to beat a dead horse (or a dead cat in this case), but let me propose a modified cat box experiment that, I think, is a legitimate test of the quantum theory as it relates to the macro-classical world. There are three principles that will guide the experiment.

1) If a quantum theory makes a prediction about the macro world and that prediction does not comport with observation in the macro world, either the quantum theory is not true or it can make no accurate prediction about the macro world because quantum laws are different from macro-matter laws.
2) If *any* outcome of an experiment supports a prediction, then the hypothesis is unfalsifiable and therefore unscientific.
3) The *mantra* one hears repeated frequently in physics culture is that the quantum level operates on different laws from the macro world, and, since our senses and mind have evolved to survive in the macro world, we cannot understand quantum reality with the rational mind. If this is true, and the laws of the micro world are different from the laws in the macro world, then predictions derived from the quantum world have no validity in the macro world where we live. Thus, a quantum probability could not result in a macro split of the universe so that each quantum probability would be true. However, if we are made of quantum particles, there must be some continuity between the micro-world and the macro-world.

My proposed experiment would be a modified Schrodinger cat box set up, which would have the usual quantum trigger (based on a 50/50 probability of a radioactive particle being emitted within a certain length of time) and the poison gas vial which would be broken if a radioactive particle is emitted. Additional elements would include the following.

1. The cat box would contain a video camera with a timer so the cat could be observed at all times. PETA or ASPCA would probably sue me for using a cat, but any form of life such as an insect would serve the purpose. There could be other monitors such as an O_2/CO_2 sensor.
2) Run the experiment 100 times.
3) Prediction is that the cat would be observed to be alive approximately 50 times and dead approximately 50 times. There would be no statistically significant difference in a large enough number of runs.
4) The cat would never be observed to be both dead and alive, nor would there be any indirect

evidence of such a state.

5) The video camera with timer would indicate the time when the poison gas was released and approximately the time when the cat died. If the cat died, *rigor mortis* would also indicate that the cat was dead before the observation was made, thus indicating that the observation of the experimenter had nothing to do with the cat's status.

6) John Wheeler would probably argue, based on the participatory anthropic principle, that the observation of the cat's death would, *ex post facto*, determine the cat's fate by going back in time and creating the history of the cat dying. This principle is unfalsifiable and therefore unscientific; because, whatever the outcome of an experiment, one could always say that it was created by one's consciousness going back in time.

7) Some might also argue that the video camera itself is a form of consciousness and its observation collapses the wave and determines the state of the cat. However, there is no evidence that mechanical devices have consciousness.

8) Wigner would probably argue that his friend, who is fond of the cat and whom Wigner left in charge of the lab, would have collapsed the wave when the friend observed the cat. However, the friend has become entangled in the quantum probability wave, so for Wigner, who has not yet made an observation, the cat would still be dead-and-alive until Wigner sees his friend's face and observes whether his friend is happy or sad about the fate of his beloved cat.

Well, this makes for a cute and fanciful story, but again, it is unscientific mysticism. When the friend makes the observation, the quantum probability wave has been collapsed in the Copenhagen interpretation. Now for Wigner, the probability ceases to be quantum and enters the macro world, so that now the probability is whether his friend is happy or sad (a macro probability). Again, this thought experiment invokes a contradiction – the cat is either dead or alive for the friend, but both dead and alive for Wigner. This is a personal relativity that even Einstein would probably reject.

9) There would be no observation that would support the Many World's theory because there is no direct or indirect evidence that the universe splits to enable each possibility to be true. The leap from a quantum probability to the whole universe splitting is a quantum leap indeed.

10) Repeat the trials 100 times, this time by flipping an evenly-balanced coin, and there will be no statistically-significant difference in the results obtained by using a macro-object (the coin) and a random quantum event (radioactive emission), thus suggesting that the quantum world does not (at least in this instance) operate on different statistical laws than the macro world.

As a post script to the contention by quantum physicists that our senses and mind are adapted to the macro-classical world and not to the quantum world which operates on different laws, let's consider our primary sense, i.e., sight. Our eyes are adapted to the quintessential quantum phenomenon – light. Now, one might argue that light illuminates macro-objects and enables us to see only into the Classical world where Newton's laws apply. However, the human visual mechanism identifies objects by the frequencies of light that they absorb and reflect, so the human eye can, in that sense, perceived differences in quantum phenomena. This observation may not refute the concept that the quantum world is different from the classical, but it does show that our vision is a combination of the quantum world interacting with the macro-world.

Quantum Cats have only Two Lives – Maybe

QUANTUM QUANDARIES AND QUACKERY — Quantizing the Unquantizable

Here are the choices for the poor quantum cat caught on the horns of a superposition.
Copenhagen: The cat is dead and alive until you look at it and kill it or make it alive.
Many Worlds: The cat remains dead and alive because the universe splits and thus splits the cat so that one cat is alive in one universe, and the other cat is dead in another universe.
Zero Worlds: The cat is neither dead nor alive because the cat is not real since the universe is not real.
EPR: There is no quantum cat. All cats are classical and live till they die.
Bohmian Mechanics: The cat is either dead or alive because all cats, like all particles, are entangled in a seamless whole.

Evolution involves a superposition: Since mutations in DNA are said to be driven by widespread radiation, the process of evolution is driven by quantum probability and uncertainty. Hence before radiation hits a gamete (sex cell) causing a mutation which either produces a higher chance of survival and reproduction OR an abnormality which may cause death for the individual receiving the mutation, that individual exists and doesn't exist and is dead and alive, until the radiation hits the gene. Moreover, since a new species being born depends on numerous such mutations, that new species exists and doesn't exist until the final mutation makes it a new species, and it is observed by another conscious being.

CHAPTER 5: THE STANDARD MODEL OF PARTICLE PHYSICS

When I took physics and chemistry in the 60s, the only subatomic particles mentioned were the proton, neutron and the electron, and we were told that these were the basic, indivisible building blocks of the universe. Even though sub-subatomic particles were known at the time, they had not made their way into the textbooks. During that period and shortly afterward, a zoo of subatomic particles was discovered by smashing larger particles in accelerators to get ever smaller units of nature. So many new particles were discovered that someone said that the Nobel Prize should go to the physicist who did not discover a new particle. This plethora of particles began to look like botany with its proliferation of species, so the Standard Model was designed to reduce this mélange of particles into a simpler model – thus the Standard Model was born. The following is a catalog of particles and their essential features.

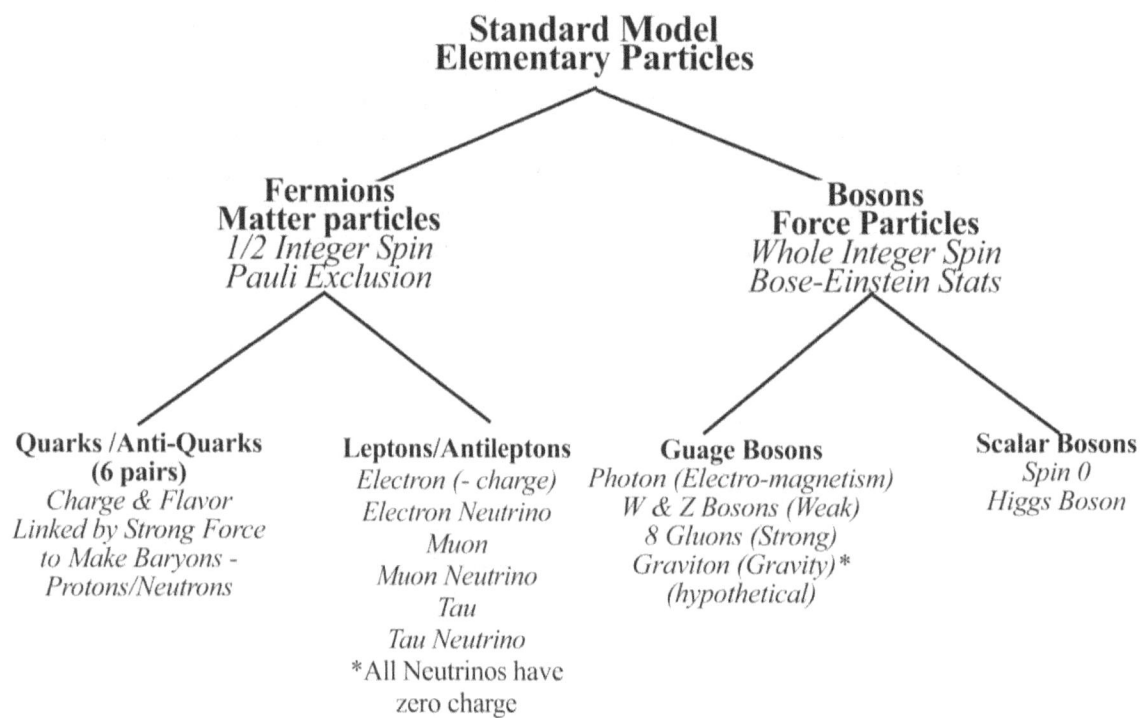

The following is my critique of the Standard Model based on linguistic analysis, logic and empiricism – the stuff of the scientific method.

Spin Direction and Magnetism: One of the most counterintuitive aspects of the Standard Model is spin. Of course, physicists cannot directly observe particles spinning, but spin is inferred from the fact that electrons supposedly act as tiny magnets. Since the earth is a spinning magnet and has a North and South Pole, the electron must spin since it, too, is a dipole having a North and South Pole. However, unlike the earth which spins one time to return to its original position (1 integer spin), electrons and other fermions (such as quarks), are said to have a ½ integer spin. In this Zen-type koan, a fermion particle must spin twice to return to its original position. This concept defies all common-sense logic and perhaps comes out of Pythagorean-type mathematics rather than direct observation or inference from observation. Particles with half-integer spin are said to follow the progression of ½, 3/2, 5/2, etc. A macro analogy used to illustrate how something must spin twice to return to its original state is the Mobius strip. An ant crawling around the strip must make two revolutions around the strip to return to its starting point. This is true because the Mobius strip is twisted so that the inside of the paper or other medium interfaces with the outside of the medium. Therefore, the ant must crawl around the inside and outside of the strip to return to its starting point. However, analogies are used to illustrate rather than argue a point, and the physical similarity between a Mobius strip and a spinning electron is probably close to nil.

Moreover, some physicists contend that the electron does not really spin, but that its bipolar personality is due to some intrinsic quality, i.e., built into it. If the electron or fermion does not spin, then this would seem to throw into doubt the whole integer and half-integer spin business. The bipolar nature of the electron brings up the concept of the unification of electricity and magnetism. It is true that electricity can be created from magnetism and magnetism from electricity, but there are some differences between the two. If the electron is the carrier of the electrical force, then it seems that it would also be the carrier of the magnetic force, and thus the carrier of the electro-magnetic force. Of course, we are told that the photon is the carrier of the electro-magnetic force, but, at the same time, electricity is said to be the flow of electrons. However light and electro-magnetic radiation stemming from electrons is carried by the photon. Be that as it may, some physicists claim that the electrons don't really flow, but bump into each other creating a chain reaction in the form of a wave. The question is, if the electron is the carrier of electro-magnetic force and it is negative in charge, how is it that it creates a north and south pole, which is also conceived as positive and negative. For example, the North Pole of the earth is pictured as negative and the South Pole is conceived as positive. Since, one thinks of positive charges as being associated with the proton and negative charges with the electron, one might think that the negative pole would be associated with a preponderance of free electrons and the positive pole as being associated with a preponderance of protons or some polarization of the atoms within the magnet. Since a magnet can be broken and still display magnetism, it would seem that all the atoms in the magnet are polarized. Nonetheless, if polarization involves some consistent alignment of electrons and protons in individual atoms, it is difficult to see how the electron can display both positive and negative properties – although some would say that the north-south of a magnet is not the same as the positive-negative of electricity. The following table illustrates this conundrum.

	Magnetism	
Electricity	North Pole (Negative)	South Pole (Positive)
Positive (Proton)		x
Negative (Electron)	x	

The table shows that the North Pole, like the electron, is associated with negative charge, and the South Pole, like the proton, is associated with positive charge. Now since polarity can change, it would be better to designate the poles of a magnet as either positive or negative. So how does the spinning electron, which is negative, create both charges or perhaps both poles? Perhaps electricity and magnetism, while unified at a deeper structural level, are different in their manifestations. J.J. Thompson's experiment involving placing a horseshoe magnet over a cathode ray tube shows that the cathode rays (stream of electrons) is deflected in opposite direction if the north and south poles are reversed in their placement (Maglab 2014). This attraction-repulsion indicates that the positive-negative property of electricity is correlated to the positive-negative (north-south) property of magnetism. This was, after all, the experiment that enabled J.J. Thompson to identify the electron as a *negatively* charged particle by the way it was deflected by magnetism.

However, it seems that the Stern-Gerlach experiment contradicts my notion that an electron can have only negative electrical charge and not positive and negative magnetism. Recall that we are told that the electron as a magnet can be oriented with the poles in any direction. The electron can have the north or south pole pointed up or down, horizontally or anything in-between. However Stern-Gerlach did not use a cathode ray (stream of electrons) in their experiment. They used silver atoms to shoot between magnetic poles with the south pole in the up position and the north pole in the down position. Silver atoms have 47 electrons (Hughes 1999), and according to the Pauli Exclusion Principle, no two electrons can occupy the same orbital if they have the same quantum numbers. In other words, electrons in the same orbital must have opposite spin directions. For example, if one electron is spin-up, then the other member of the pair must be spin-down. Hence all the electrons of silver are paired up and would cancel each other's spin except the 47th one which is unpaired. Now that 47th electron might be attracted to the positive pole of the magnet thus deflecting it in that direction and repelled by the negative pole of the magnet. Thus, the poles of a magnet would be correlated to the negative charge of the electron.

So, the question becomes: was the whole atom of silver spinning or were only the electrons spinning. To use the planetary model of the atom, perhaps the nucleus spins like the sun and the electrons orbit the nucleus and spin like the planets - and perhaps the whole atom is in spin like most other things in the universe. It is not necessary to assume that all the polarization of the atom is due to electron spin because, as a negatively charged particle, the electron would seem to be unable to generate a positive charge – again assuming that the north-south poles of a magnet are correlated to electric charge. It is difficult to visualize the polarization of an atom with negative charges on one side and positive charge on the other side. This picture contradicts the planetary image that electrons are constantly orbiting the nucleus, so their negative presence would be manifested in all directions around the atom. Although I believe that ultimately these paradoxical pictures emanating from quantum mechanics can be resolved without resorting to mysticism and counterintuitive theories, for now we must accept that paradoxes result from incomplete understanding, not uncertainty in nature.

In the Stern-Gerlach experiment, after the stream of silver atoms are shot through the vertically arranged poles of a magnet, the up-spin atoms are then passed between the poles of a horizontally-arranged magnet, and the stream of atoms split horizontally (right and left) suggesting that they have a horizontal polarization as well as a vertical one. This result was surprising because the experimenters thought that if atoms have an up or down spin orientation, they should not have a right and left spin orientation also and they would continue in a straight line. Then the same stream of silver atoms passes through vertically organized magnets again and the stream again splits between an up and down orientation. This was really surprising because the atoms had already been sorted as having an up-spin orientation. What this shows me is that the magnetic poles of an atom can switch. Most of us as kids have had the experience of playing with magnets and having them reverse polarity so that the sides of the magnets that once attracted each other, now repel each other. Furthermore, the magnetic earth is said to reverse polarity every thousand years or so. Some would attribute this apparently random reversal of polarity to the Heisenberg uncertainty principle, but there might be a predictable cause for the switching of polarity. Also, as kids we learned that if you place magnets at some angle to each other, they will twirl so that opposite poles will attract. This twirling could be what is happening in the second pass when the up-spin atoms are run through a horizontally-arranged magnet. This realignment might then affect the twirling of the third pass when the atoms are again run though a vertically arranged magnet.

Now, the above analysis of spin/magnetism is to question established interpretations and offer an alternative view. The analysis certainly shows that there are still many unanswered questions about subatomic spin and magnetism which manifests itself at the micro and macro level. Certainly, if positive and negative electric charges are correlated to north and south poles of magnets, then electro-magnetism would seem even more unified. One difference between electricity and magnetism is that, in electricity, positive and negative charges can exist independently (electrons can exist apart from protons), but the poles of a magnet must occur together. However, some theories predict monopole magnets, but no such magnet has ever been observed.

Electricity and Magnetism related to Electro-chemical Processes

Most of us have had the experience in chemistry class of creating a current with a voltaic cell which involves two metals (electrodes) in a salt solution. In this process, we create a simple battery. The voltaic cell involves stealing electrons from one electrode and giving them to the other electrode in an oxidation-reduction reaction. While the electrons are flowing from one electrode to the other, we can attach a light or other detector and demonstrate that we are generating electricity. Electrolysis involves the opposite process. By running an electric current from one electrode to the other, we can oxidize one electrode (steal electrons) while reducing the other electrode (adding electrons to neutralize it and making a solid from dissolved ions). The following is a table showing the two processes.

	Galvanic/Voltaic		Electrolytic	
	Lose e- (Ox)	**Gain e-(Red)**	**Lose e-(Ox)**	**Gain e-(Red)**
Anode		Positive to Neutral	Becomes More Positive	
Cathode	Becomes More Positive			Positive to Neutral

My proposed experiment is to use a similar set-up to see if a magnet behaves in a way similar to producing electricity from chemistry. If a horseshoe magnet is placed in a saline solution, will electricity flow from one pole to the other and, if so, in which direction, will it flow. Furthermore, if we shoot cathode rays (electrons) in the solution, will they gravitate to one pole or the other? Or if we run electric current into the solution, will the current flow to one pole and be repelled by the other? This is very similar to J.J. Thompson's experiment except that he shot electrons in a near vacuum tube and tested the effect of a magnet placed in alternating north-south positions on the deflection of the stream of electrons. The essential question is whether the electrodes of an electro-chemical reaction are the same as the poles of a magnet. After all, in the electro-lytic process the electrodes (cathode and anode) are electro-magnets. The question then becomes: "Does an electro-magnet behave the same way as a regular magnet in its attraction-repulsion?"

Gravity, the weak link in the quantum chain

Einstein described gravity as "Nature's best kept secret." Gravity eluded Einstein in his search for a unified field theory, and it has eluded quantum physicists who are seeking particles to carry all the known forces of nature. While there are particles identified as the carriers of electro-magnetism, the strong force and the weak force, no such quantum particle has been found for the pervasive force of gravity. In the parlance of quantum physics, gravity has not yet been *quantized* even though the wizards of quantum loop gravity have tried.

Despite not having yet found a gravity particle or graviton, physicists believe they have found gravity waves predicted by General Relativity. Some physicists claim that gravity is the weakest force of the four, and therefore, the elusiveness of the gravity particle (the hypothetical graviton) may be due to the fact that it is a small particle having an incredibly short wavelength which is said to be 1/10,000 of the width of a proton (Barish and Weiss 1999: p. 44). However, even though gravity may appear weak in the short range, it has the farthest reach of any known force. The strong force is said to be the strongest force, but it operates only over very short distances. The weak force also is a short-range force. Electricity, although said to be stronger than gravity, does not have a range as long as gravity's range because electricity gets neutralized as it travels through space and encounters positive particles. Nevertheless, the photon which is the carrier of the electro-magnetic force has a reach as far as gravity and exerts radiation pressure in the form of heat, but it could hardly be as forceful as gravity in the long run. It is interesting that Sakharov and Brandenburg have hypothesized that the photon is also the carrier of gravity and its force depends on the amount of radiation pressure in a given area. That difference in pressure is supposed to determine the motion of heavenly bodies.

It seems the claim that gravity is the weakest force needs some qualification. Admittedly, it takes a lot of mass to create strong gravity, but when it reaches a critical level, gravity exerts an extremely long-range influence on distant objects. When a star the size of our sun can hold the planetoid Pluto into an orbit, it seems that gravity is no puny force. Since our sun's diameter is 864,575 miles and Pluto is 4.67 billion miles from the sun, the sun exerts an orbit-compelling gravitational force on Pluto that is 4,245 times greater than the sun's diameter. If one compares the diameter of the sun to the farthest known object in our solar system, a planet-like body called

QUANTUM QUANDARIES AND QUACKERY — Standard Model

Sedna, which is 8 billion miles from the sun, then the sun exerts an orbit-compelling force that is 9,253 times greater than its diameter. That does not seem like a weak force to yours truly. It appears that there is an inverse relationship here. The strongest force at short range is the weakest force long range, and the weakest force short range is the strongest force long range. Furthermore, if one considers neutron stars in which gravity is said to crush electrons into the nucleus of all its atoms so that all its protons become neutrons, then such a force is not to be underestimated. Likewise, if one considers that a black hole, the gravity monster, is so strong that nothing, not even light, can escape its grasp, then it would seem that gravity (based on ultimately dense mass) is the strongest force in the universe – short range and long range. So, let's don't short change gravity as the weakling in the pantheon of forces – the strength of a force seems to depend on how one measures it.

The following table compares the strength of the four forces short range and long range.

Forces	Short range	Long range
Strong Nuclear Force	Strongest	May be Weakest
Weak Nuclear Force	Strong	Weak
Electro-Magnetism	Strong	Strong
Gravity	Weakest	Strongest

Quite often, a physicist will illustrate the strength of electro-magnetism over gravity by dropping an object from a multi-story building and observe the object bouncing back when it hits the earth. The conclusion of the experiment is that the electro-magnetic bonds that hold the earth together resists the fall of the object under the influence of gravity. However, the fall of the object due to gravity is actually going up against three forces, not one. The three forces are:

1) **The strong force** that holds the nucleus of atoms together. Rutherford discover that when alpha particles are shot through a thin sheet of gold foil, occasionally one bounces back because it hits the nucleus. So, a falling object encounters this force when it hits the ground.

2) **Electro-magnetism**: The positive nucleus containing protons is what attracts negative electrons and holds them in an orbit around the nucleus. However, electrons can be stripped away from their orbits and high gravity can crush the electrons from their orbits into the nucleus as in a neutron star.

3) **Gravity**: The other force that holds this terrestrial ball called earth together and resists the falling object is gravity itself. So, in this scenario, we have gravity competing with gravity plus the two other forces.

What this experiment leaves out is that some objects will break and shatter when they hit the ground, meaning that the electro-magnetic bonds that holds atoms and molecules together have been broken. Here we would conclude that gravity plus the other forces have exerted power over the falling object. Of course, that gravity has been generated by a tremendous amount of mass, i.e., the mass of the earth. However, an experiment performed on the space station shows that objects in a solution are attracted to each other, supposedly because of micro-gravity. Again, our conclusion is that gravity, while it may be weak in the short-range, is the strongest of the forces in the long range.

Supersymmetry (SUSY)

Before there is supersymmetry, there must be symmetry. In physics, symmetry means something that looks the same no matter which way one looks at it. For example, a sphere looks the same, no matter which way you roll it even if you are looking at different points on the sphere. The concept of supersymmetry is somewhat analogous to the concept of anti-matter. True to our dualistic cognitive structure, it was predicted that for every matter particle, there must be an anti-matter particle of opposite charge. This idea led Dirac to work out the mathematics from Special Relativity to predict that such anti-particles exist. Later, physicists found what they were looking for in a particle collider - antimatter. SUSY now proposes another duality -- for every matter particle (fermion) in the Standard Model, there should be a force particle (boson) that is symmetrical or the same in all its properties such as spin, mass, charge, etc. For such superpartners to be the same, the equations that describe the super particles must be rearrangeable and give the same result (Lincoln 2013).

The reason that there is a need for supersymmetry is that the symmetry of the particles in the Standard Model is supposedly broken. Hence the particles in the Standard Model are incomplete, even though all matter particles have anti-matter complements. To enable the Standard Model to explain more quantum phenomenon, each particle needs a *superpartner* in order to account for the Higgs field which gives mass to particles, to unify the forces of nature, and to explain gravity and dark matter. It would seem at first blush that the known fermions and bosons are partners to each other already. The photon (boson), which carries the electromagnetic force, is connected to the electron; the strong force is carried by the gluon and binds quarks into protons and neutrons in the nucleus; the weak force is carried by the W and Z bosons, and hypothetically gravity is carried by the graviton. However, these force particles are not symmetrical to their matter particles. For example, the photon is said not to have mass, but its larger, slower cousin, the electron, does have mass. Also, the other force particles are not symmetrical to the matter particles they carry.

To apply critical thinking to supersymmetry, these points should be considered:
1) No supersymmetric particle has ever been detected in a particle collider despite extensive searching. Nonetheless, there have been 10,000 scientific papers written on the subject (Lincoln 2013).
2) Fermions (matter particles) have ½ integer spin and bosons (force particles) have whole integer spin. If spin is the key distinction between fermions and bosons, then how can matter and force particles ever be symmetrical – one will always have half integer spin and the other whole integer spin?
3) Some bosons are massless, so it would seem impossible for their matter superpartners to be massless and still be a matter particle. For example, how would the photon or gluon, which are said to be massless, find corresponding matter superpartners, which would, of course, have mass?
4) Supersymmetry requires at least one extra dimension – the quantum dimension. As indicated previously regarding String Theory, extra dimensions (beyond the three we experience) involve pure speculation and none have ever been observed. Moreover, to my knowledge, there is no indirect evidence for them. Extra dimensions are invoked to make mathematics work.
5) Recall that Einstein said that "imagination is more important than knowledge"; however, the imagination can create things that have no counterpart in physical reality. Perhaps that is the case

with superpartners existing for the known particles in the Standard Theory, and nature apparently does not provide such partners.

6) The theory is primarily mathematical, and mathematical metaphysics does not always represent reality. To wit, Supersymmetry is sometimes studied mathematically for its intrinsic properties. This is because it describes complex fields satisfying a property known as holomorphy, which allows holomorphic quantities to be exactly computed. This makes supersymmetric models useful "toy models" of more realistic theories. A prime example of this has been the demonstration of S-duality in four-dimensional gauge theories that interchanges particles and monopoles (Krasnitz 2002: p. 91).

Given this dubious state of affairs for supersymmetry theory, it appears that it may go the way of String Theory, which has also jumped on the supersymmetry bandwagon with its own *Superstring* theory. However, the two theories, which seem tied to a common fate, are languishing for lack of evidence and hanging on by a thread. Furthermore, the extra dimensions of hyperspace are in serious doubt. JHM Alok says:

Some particle physicists became disappointed by the lack of experimental verification of supersymmetry, and some have already discarded it; Jon Butterworth at University College London said that we had no sign of supersymmetry, even in higher energy region, excluding the superpartners of the top quark up to a few TeV. Ben Allanach at the University of Cambridge states that if we do not discover any new particles in the next trial at the LHC, then we can say it is unlikely to discover supersymmetry at CERN in the foreseeable future. (Alok 2013).

Moreover, one of the former enthusiastic supporters of supersymmetry theory in the 1980s, Mikhail Shifman, urged the theoretical physics community to search for new ideas and acknowledge the reality of the failure of supersymmetry theory in order to help the new generation of theoreticians to avoid useless work and becoming a lost generation (2012).

> Perhaps SUSY is based upon a wrong assumption about nature, i.e., that all matter particles need a force particle (some of which are massless) to carry them. To begin with, the idea of a massless particle is a self-contradiction – a particle of energy (force) or a particle of matter has mass. Second, it may be that matter particles have their own intrinsic force to motivate their behavior and that the matter particle would be identical or symmetrical to its force partner is a self-contradiction. If they were identical, then there would be no need to make a distinction between matter and force – they would be one and the same. However, one might argue that symmetrical and identical are not the same concept, but if the properties and equations are the same, then perhaps the two concepts are the same. There may be something inherently wrong in the assumption and semantics of supersymmetry.

CHAPTER 6: UNIFICATION THEORIES

To unify the known forces of nature has been a dream in physics from the time of Einstein who worked unsuccessfully in the waning years of his life to unify electromagnetism and gravity in the manner that Maxwell had united electricity and magnetism with one set of equations. Now modern physicists aspire to unite the strong and weak forces with electromagnetism in what is called GUT (Grand Unification Theory). To unite GUT with gravity, the rogue force, is the ultimate dream referred to by the acronym TOE (Theory of Everything). Thus far, physicists seem to agree that quantum mechanics has been united with electromagnetism (QED) and the weak nuclear force has been united with electromagnetism (Electroweak unification). However, as Einstein said, "Gravity is nature's best kept secret", and it has been most resistant to any kind of unification. Much of this unification theory is based upon the speculative notion that all the forces were one at the beginning of the Big Bang when there was a singularity. As the universe cooled and the symmetry between the forces was broken, the forces began to differentiate and become what they are today, serving specialized functions in making the universe work and for life to come into existence.

QED (Quantum Electrodynamics'): The Normalization of Renormalization

Unifying electromagnetism and quantum mechanics

There were three physicists who created QED, but Richard Feynman is given primary credit for it because of his simplification of complex equations in the form of Feynman diagrams. Each Feynman diagram is a picture of an equation which describes the interaction of electrons and photons in various combinations. Whereas classical electromagnetic theory speaks of electric and magnetic fields as lines of force (which are pictured as continuous), quantum theory converts the fields to particle interactions which are discrete or quantum. The particle that mediates between electrons and creates the repulsive force is, of course, the photon. Now, to achieve this unification, a very complex equation was written, which no one has ever been able to solve because the equation involved infinities. Of course, infinities are not amenable to real-world mathematical computation and do not, in a finite context, yield realistic physical results – since something that is finite cannot be infinite. Now to achieve the impossible dream and deal with infinities, **renormalization**, based on perturbation theory, was created. The idea behind perturbation theory is to create a simpler form of a complex equation so that it can be solved. However, the simpler equation (in this case, the renormalization equation) is an approximation of the original, complex equation and sacrifices accuracy for solvability. The key to renormalization is to eliminate infinities.

Now, getting rid of infinities is all well and good, until one starts to solve the *magnetic moment* (strength) of an electron. The electron magnetic moment is designated as 1, but perhaps because of the presence of the photon cloud around the electron, it is a little larger than 1. A physicist by the interesting name of Polykarp Kusch calculated that the magnetic moment as 1.00119. However, since Kusch made the prediction, physicists have improved their techniques and have made a more accurate prediction. Below, we see a very slight deviation between the mathematical prediction and the experimental value for the magnetic moment of an electron. As I understand it, the measurement of this magnetic moment is based upon the amount of precession (wobble) in the electron spin. It would seem that the stronger the force, the less wobble there would be, and, conversely, the weaker the force, the greater the wobble – as can be seen in a spinning top.

Predicted Magnetic Moment of electron = 1.001159652182
Measured Magnetic Moment of electron = 1.001159652181

As can be seen, the deviation between prediction and measurement occurs in the 12th decimal place and is only 1 part in a trillion (The foregoing discussion on QED is based on Donald Lincoln's lectures on QED on Fermi Lab YouTube site). Lincoln says that the close agreement between the prediction from the equation and measured result shows that QED is an extremely accurate theory and demonstrates the power of QED to predict the behavior of electrons. While this apparent accuracy is extremely impressive, Lincoln seems to forget that renormalization, based on perturbation theory, uses, not the original, more accurate formula involving infinities, but an APPROXIMATE, less accurate formula which simplifies the equation by eliminating infinities and making it solvable. If the renormalization formula is used to calculate the magnetic moment, it is reasonable for the outsider dissident to question this extreme accuracy. Moreover, renormalization was initially viewed as a suspect provisional procedure even by some of its originators; however, renormalization eventually was embraced as an important, self-consistent, actual mechanism in several fields of physics and mathematics. Nonetheless, here is what the founder, Richard Feynman (1990), who shared a Nobel Prize for the theory's development, says about it:

The shell game that we play ... is technically called 'renormalization'. But no matter how clever the word, it is still what I would call a dippy process! Having to resort to such hocus-pocus has prevented us from proving that the theory of quantum electrodynamics is mathematically self-consistent. It's surprising that the theory still hasn't been proved self-consistent one way or the other by now; I suspect that renormalization is not mathematically legitimate (p. 128).

Furthermore, Paul Dirac (who predicted anti-matter thus uniting Special Relativity with Quantum Mechanics) criticized the theory as follows: "Sensible mathematics involves neglecting a quantity when it is small – not neglecting it just because it is infinitely great and you do not want it! (Dirac 1963: p. 53). I agree with Dirac, that eliminating small quantities may be acceptable practice, but eliminating infinities is a violation of both mathematical and scientific principles.

Furthermore, Gribbin (2009: Kindle Book Location 2567) says that the way infinities were eliminated from the renormalization equations was to divide one infinity by another to produce the number 1. The notion of dividing infinity by anything, including infinity, presents serious problems in logic. To begin with, the idea of an infinite *number* is an oxymoron. Infinity is not a number. A number always indicates a finite amount of something; therefore, to speak of an infinite number is tantamount to saying a finite infinity – another Zen koan. Rather than speak of an infinite number, one should simply say "infinity". To divide infinity by infinity is to say there are two infinities. If one is talking about quantities of the same thing, then there cannot be two separate infinities of that thing. One infinity rules out the possibility of a second infinity because, the first alleged infinity would not be infinite if one could add to it or produce another just like it. The idea of reducing an infinity to the number 1 is, for lack of a better word, ludicrous. Again, we find that much of modern physics is numerology with little basis in the real physical world.

The problem seems to stem from *deductive bias* which motivates physicists to make the math fit the theory instead of making the math conform to the real physical world. Since, over time, renormalization has gradually become accepted in the physics community (as physicists have become comfortable with ignoring infinities), I call the process the "normalization of renormalization." This process is reminiscent of "Planck's constant" which Max Planck created as a fix for the Rayleigh-Jeans equations which predict infinite energy emitted from the Black Body (known as the "ultraviolet catastrophe"). Although Planck regarded his constant (which resolved the infinities by quantization) as a mathematical trick, which just happened to work, later physicists came to accept it as a reality and use it extensively in quantum equations. It is also reminiscent of Ptolemy's epicycles inserted into the motion of planets which was a fix to make the geocentric theory work. What it amounts to is shielding a theory from falsifiability.

Electroweak Unification

The electroweak unification (Nave 2016) was made possible by combining the Big Bang theory with the discovery in 1983 of W and Z particles whose existence had been predicted by Weinberg, Salam, and Glashow in 1979 for which they won a Nobel Prize. The unification of electromagnetism and the weak force involves the identification of the photon, carrier of electromagnetism, with the W and Z particles associated with the weak force which contributes to nuclear disintegration and the formation of lighter elements from nuclear splitting. The writers of the hyperphysics article say that:

The theory suggests that at very high temperatures where the equilibrium kT energies are in excess of 100 GeV, these particles are essentially identical and the weak and electromagnetic interactions were manifestations of a single force (Nave 2016).

This analysis seems to suggest that electromagnetism is not unified with the weak force in the present universe, but these forces were united in the early history of the universe. This theory seems highly speculative and has several challenges.

1~The W and Z particles are very massive, but the photon is said to be massless. How could electromagnetism and the weak force be manifestations of the same underlying reality with such differences in their carrier particles? As discussed previously, the photon cannot be a massless particle, but since it is very light, it seems to be no match for W and Z particles. The writers of this article indicate that "the difference in masses can be attributed to spontaneous symmetry breaking as the hot universe cooled." The difference would be absolute if the W and Z are very massive and the photon has zero mass.

2~"The symmetry-breaking mechanism is called a Higgs field, and requires a new boson, the Higgs boson to mediate it" (Nave 2016). As indicated previously the Higgs particle is very speculative and some physicists acknowledge that the claim of its discovery is not very certain.

3~Much of Big Bang theory is also very speculative and many of its claims have been disputed by reputable scientists.

Overall, this alleged unification seems to be built upon a shaky foundation of questionable assumptions. If the breaking of this unification occurred eons ago when the early hot universe began to cool, then it is not a unification in the same sense that electricity and magnetism are unified in the present world. This seems to be a weak unification indeed.

The zero-infinity problem with Standard Model

John Gribbin (2009) describes the perennial problem with zero and infinity in the Standard Model, and I would say that Modern Physics in general has this problem of logic.

(An)…embarrassing feature of the standard model of particle physics based on the idea of electrons, quarks and other entities as point-like particles. If the particles literally existed as mathematical points with zero spatial size (zero dimension), the theories were plagued by unwelcome infinities. In a simple example, think of the inverse square law of the electric force. The force between two charged particles is proportional to a 1 divided by the square of the distance between them. If a charged particle had zero volume, then the force acting on the particle itself (its self-interaction) would be infinite: 1 divided by zero squared. Things like electrons ought to explode as a result (p. 145).

The fallacies are apparent in this description:

1>A particle with mass or energy cannot have zero dimensions. A point is a geometric concept of zero dimension which is the starting or ending place. Particles which combine to make up macro matter cannot have zero dimensions or zero mass because no amount of nothingness (zero) can make something. Here again, physicists mix a pure metaphysical concept with a physical reality.

2>The idea that 1 divided by zero yields infinity leads to illogical conclusions. Since division and multiplication are reversible, if $1/0$ = infinity, then infinity times $0 = 1$. Both the division and multiplication of zero and infinity involve absurd answers.

CHAPTER 7: CONCLUSIONS REGARDING QUANTUM PHYSICS

As stated before, an important principle for evaluating theory is that when one takes a theory to its logical conclusion and it reduces to absurdity, then something is probably wrong with the theory. Accordingly, when quantum mechanics is combined with relativity, and an absurd answer emerges, then something must be wrong with one or both theories. It is in such dilemmas as these, that physicists resort to constants, epicycle-type patches and outright mysticism. I think that there is something wrong with both theories (relativity and quantum mechanics) because of faulty assumptions and premises, and there needs to be a re-examination of both theories at the foundational level. Here are some of the catch-22 dilemmas in which Cox and Forshaw (2014) get themselves entangled when quantum mechanics is merged with relativity.

The energy stored up within 1 cubic metre of empty space as a result of quark and gluon condensation is a staggering 1035 joules, and the energy due to Higgs condensation is 100 times larger than this. Together, that's the total amount of energy our Sun produces in 1,000 years. To be precise, this is 'negative' energy, because the vacuum is lower in energy than a Universe containing no particles at all. The negative energy arises because of the binding energy associated with the formation of the condensates.

Here we are treated to the fallacy that negative energy is less than no energy at all (zero energy). Scientific realism would dictate that you either have energy or you don't. Repulsive energy compared to attractive energy is not less than zero energy. Negative numbers used in this sense are purely imaginary. Cox/Forshaw continue:

What is mysterious, however, is that such a large and negative energy density in every square metre of empty space should, if taken at face value, generate a devastating expansion of the Universe such that no stars or people would ever form. The Universe would literally have blown itself apart moments after the Big Bang. This is what happens if we take the predictions for vacuum condensation from particle physics and plug them directly into Einstein's equations for gravity, applied to the Universe at large. This heinous conundrum goes by the name of the cosmological constant problem and it remains one of the central problems in fundamental physics (Kindle Book: Location 2990-3001).

> **To put Cox and Forshaw's argument in a nutshell:**
> 1) The negative (repulsive) energy in one cubic meter of *empty* space is equal to the amount of energy emitted by the sun in 1,000 years.
> 2) Since this energy is negative (perhaps dark energy), it would cause space to expand rapidly.
> 3) Such rapid expansion of space would not have allowed stars and galaxies to coagulate after the Big Bang because the repulsive force would have been much more powerful than gravity.
> 4) Thus Einstein's theory of gravity in general relativity (even with his balancing act of the cosmological constant) does not jive with this quantum interpretation of space-expanding energy.

In the above statement, we see what happens when an imaginary concept such as negative energy is invoked. This assumption of negative energy (less than zero energy) leads to absurd conclusions that are obviously unempirical. It seems to me that Cox and Forshaw's argument is self-refuting. It is ironic that, as previously stated, Sakharov and Brandenburg thought that repulsive radiation pressure in space is what creates gravity, thus uniting quantum mechanics, electro-magnetism and gravity. Here Cox and Forshaw are saying the opposite. They are saying that these two forces are contradictory and cannot be united to make quantum theory harmonize with relativity.

In a rare instance of humility, Cox and Forshaw are compelled to admit:

Certainly it suggests that we should be very careful before claiming to really understand the nature of the vacuum and/or gravity. There is something absolutely fundamental that we do not yet understand (Kindle Book: Location 3001).

Why this realization would not cause them to question more of the foundational assumptions of relativity and quantum mechanics is a testament to the power of authority over our minds. Despite this unusual statement of humility, they continue to tout the theories of Modern Physics with great certainty. One such statement is that since macro matter is made of quantum particles, then the universe is quantum and all levels of reality behave in accordance with the laws of quantum weirdness. Here's what they say on that score.

The laws of quantum theory replace Newton's laws and furnish a more accurate description of the world. Newton's physics emerges out of the quantum description, and it is important to realize that the situation is not 'Newton for big things and quantum for small': it is quantum all the way (Kindle Locations: 265-267).

This statement betrays a lack of understanding of the holistic principle, i.e., "the whole is greater than (or perhaps different from) the sum of the parts." This principle can be seen in the chemical properties of water, H_2O. The oxygen in water, by itself as an element, supports combustion; and the hydrogen, by itself, is highly flammable. However, when combined in this particular molecular configuration, water is non-flammable and is used to extinguish fires. So even if micro-particles behave in weird ways (which I doubt), that does not mean that the larger masses they comprise behave in the same way.

References

Abbott, B. (2007). "Microwave (WMAP) All-Sky Survey". Hayden Planetarium.

Aharonov, Yakir and Rohrlich, Daniel (2005) *Quantum Paradoxes: Quantum Theory for the Perplexed.* Wyley VCH

Aharonov, Yakir (2002) *Uncertainty.* Discovery Science Video.

Albert Einstein Site Online. Retrieved from http://www.alberteinsteinsite.com/quotes/einsteinquotes.html.

Alok, Jha, (August 6, 2013). *"One year on from the Higgs boson find, has physics hit the buffers?".* The Guardian: London.

Ananthaswamy, Anil (2017) *A classic quantum test could reveal the limits of the human mind.* New Scientist. https://www.newscientist.com/article/2131874-a-classic-quantum-test-could-reveal-the-limits-of-the-human-mind

Ashby, Neil (2002) "Relativity and the Global Positioning System." *Physics Today*, May 2002, p. 41.

Barish, Barry C. and Weiss, Rainer (October 1999). "LIGO and the Detection of Gravitational Waves". Physics Today. 52 (10).

BBC Documentary (2008) *Hawking Radiation.* Interview with Professor Bernard Carr, Queen Mary University of London. https://www.youtube.com/watch?v=S6srN4idq1E

Beckmann, Petr (1987) *Einstein Plus Two.* Boulder, CO: Golem Press.

Behe, Michael J. (1996). *Darwin's Black Box: The Biochemical Challenge to Evolution.* New York: Free Press.

Berman, Bob (Dec. 2017) "Intelligent Design" in *Astronomy* Magazine.

Bethel, Tom (2009) *Questioning Einstein: Is Relativity Necessary?* Pueblo, CO: Vales Lake Publishing.

Biever, C. (6 July 2012). "It's a boson! But we need to know if it's the Higgs". New Scientist. Retrieved 2013-01-09.

Bishop, Owen (1984) *Yardsticks of the Universe.* New York: Peter Bedrick Books.

Boesgaard, A. M. and Steigman, G. (1985). "Big Bang Nucleasynthesis: Theories and

Observations", *Ann. Rev. Astron. and Astrophys*. 23, 319.

Boswell, J. (1823). The Life of Samuel Johnson, vol. 1. London: J. Richardson & Co.

Bohm, David (1980) *Wholeness and the Implicate Order*. New York: Routledge & Kegan Paul.

Brainy Quotes: http://www.brainyquote.com

Brandenburg, John (2011) Beyond Einstein's Unified Field. Kempton, IL: Adventures Unlimited Press.

Brown, James Cooke (1975) *Loglan 1: A Logical Language,* Loglan Institute.

Bryce, Emma (2016) *Will Wind Turbines Ever be Safe for Birds.*
http://www.audubon.org/news/will-wind-turbines-ever-be-safe-birds

Bunge, Mario (2001). *Philosophy in Crisis: The Need for Reconstruction.* Amherst, New York: Prometheus Books.

Capra, Fritjof (1999) *The Tao of Physics*. Boston: Shambhala Publications.

Carrey, Jim (1995) *Ace Ventura: When Nature Calls*. Movie. Morgan Creek Productions.

Carroll, Sean (May 20, 2013) *Arrow of Time - Sixty Symbols.*
https://www.youtube.com/watch?v=9VFGuupXwng#t=235.460771

Carroll, Sean (Jan 22, 2013) *Quantum Mechanics (an embarrassment) - Sixty Symbols.*
https://www.youtube.com/watch?v=ZacggH9wB7Y

Carroll, Sean M. (2006). *C-SPAN broadcast of Cosmology at Yearly Kos Science Panel, Part 1.*

Carroll, Sean M. (2004). *Spacetime and Geometry*. Addison Wesley.

Case, Thomas. (2013) *A Short Introduction to Metaphysics*. (Kindle Locations 49-50). Didactic Press. Kindle Edition.

Caughill, Patrick (July 28, 2017) *A New Breakthrough in Quantum Computing is Set to Transform Our World.* Futurism website: https://futurism.com/a-new-breakthrough-in-quantum-computing-is-set-to-transform-our-world/

Cherenkov, Pavel A. (1934). "Visible emission of clean liquids by action of γ radiation". Doklady Akademii Nauk SSSR 2: 451. Reprinted in Selected Papers of Soviet Physicists,Usp. Fiz. Nauk 93 (1967) 385. V sbornike: Pavel Alekseyevich Čerenkov: Chelovek i Otkrytie pod redaktsiej A. N. Gorbunova i E. P. Čerenkovoj, M.,"Nauka, 1999, s. 149-153.

CERN Document Server, http://home.cern/topics/antimatter

Cervantes-Cota, J.L.; Galindo-Uribarri, S.; Smoot, G.F. (2016). "A Brief History of Gravitational Waves". *Universe* 2 (3): 22. doi:10.3390/universe2030022

Chomsky, Noam (1957) *Syntactic Structures.* The Hague: Mouton.

Clegg, Brian (2012) *Gravity: How the Weakest Force in the Universe Shaped Our Lives.* St. Martin's Press. Kindle Edition. (pp. 149-150, Kindle location 1787).

Clegg, Brian (2014). *30-Second Quantum Theory.* New York: Metro Books.

Cox, Brian (2011) *The Quantum Universe: (And Why Anything That Can Happen, Does)* Da Capo Press. Kindle Edition.

Cox, Brian and Forshaw, Jeff (2010) *Why Does E=MC².* Da Capo Press, Kindle Edition.

Davies, Paul (2003) *How to Build a Time Machine.* Penguin Publishing Group. Kindle Edition.

Davies, Paul and Gregersen, Niels Henrik (2010) *Information and the Nature of Reality: Physics to Metaphysics.* New York: Cambridge University Press.

Davies, Paul (2006) Interview in: "The Anthropic Principle" Video by IBBC Worldwide Ltd.

Del Rosso, A. (19 November 2012). "Higgs: The beginning of the exploration". CERN Bulletin. Retrieved 2013-01-09.

Deutsch, David *(1998) The Fabric of Reality: The Science of Parallel Universes.* Amazon Kindle Book.

Deutsch, Sid (2005) *Einstein's Greatest Mistake: The Abandonment of the Ether.* New York: iUniverse, Inc.

Dirac, P.A.M (May 1963) "The Evolution of the Physicist's Picture of Nature," in Scientific American, May 1963, p. 53.

Discovery Science Video (viewed 2002) *Uncertainty.*

Dyson, F.W.; Eddington, A.S.; Davidson, C.R. (1920). *"A Determination of the Deflection of Light by the Sun's Gravitational Field, from Observations Made at the Solar eclipse of May 29, 1919".* Phil. Trans. Roy. Soc. A 220 (571-581): 291–333.

Eagle, Bob, aka Dr. Physics (April 9, 2012) *Bell's Inequality.* https://www.youtube.com/watch?v=qd-tKr0LJTM

Einstein, Albert (1916) *Memorial Notice for Ernst Mach,* Physikalische Zeitschrift 17: 101-02.).

Faye, Jan (Fall 2014 Edition) "Copenhagen Interpretation of Quantum Mechanics", The Stanford Encyclopedia of Philosophy Edward N. Zalta (ed.). https://plato.stanford.edu/archives/fall2014/entries/qm-copenhagen

Feynman, Richard P (1990) *QED, The Strange Theory of Light and Matter*, Penguin, p. 128

Feynman, Richard (1965) *The Character of Physical Law.*

Folger, Tim (June, 2002) *Does the Universe Exist if We're Not Looking?* Discover Magazine.

Frank, Adam (2016). *Three Atoms per Cubic Meter*. In NPR Blogger 13.7: Cosmos and Culture. Retrieved from http://www.npr.org/2016/08/09/489361654/short-answers-to-big-questions-exploring-atoms-in-space.

Gardner, Martin (1997) Relativity, Simply Explained. Dover Publishing.

Garret, Ron (2011) *The Quantum Conspiracy: What Popularizers of QM Don't Want You to Know*. Google Tech Talk. https://www.youtube.com/watch?v=dEaecUuEqfc.

Gawiser, E.; Silk, J. (2000). "The cosmic microwave background radiation". *Physics Reports* 333–334: 245–267.

Gibbs, Phillip (1996). *"Can Special Relativity Handle Accelerations?"* The Original Usenet Physics FAQ. Retrieved 2014-07-23. http://math.ucr.edu/home/baez/physics/Relativity/SR/acceleration.html

Gingerich, Owen (1992) *The Great Copernicus Chase.* Sky Publishing Corp.http://astro.wsu.edu/worthey/astro/html/im-lab/stonehenge/stonehenge.html

Goldsmidt, Walter (1970) *Whatever Happened to Human Nature.* Lecture at Wake Forest University.

Good Reads: http://www.goodreads.com/quotes

Greene, Brian (1999) *The Elegant Universe.* New York: W.W. Norton and Company.

Greene, Brian (1999) *The Elegant Universe: Superstrings, Hidden Dimensions, and the Quest for the Ultimate Theory.* W. W. Norton & Company. Kindle Edition.

Greene, Brian (2016) *M-Theory, String Theory and the Elegant Universe.* Discovery Channel - String theory rare documentary, National geographic retrieved from Youtube: . https://www.youtube.com/watch?v=qtaAM84Kt2I.

Greene, Brian (2004) *The Fabric of the Cosmos.* New York: Random House.

Greene, Brian (2011). *The Fabric of the Cosmos:* Video by NOVA and PBS.

Greene, Brian (2014). *The Fabric of the Cosmos: What is Space?* Youtube: Rising Life Media.

Gribbin, John R. (1984) *In Search of Schrodinger's Cat.* Bantam Books.

Gribbin, John R. (2009) *In Search of the Multiverse.* John Wiley and Sons.

Gules and Sable: *Escutcheons of Science*

Gunion, John F.; Haber, Howard; Kane, Gordon; Dawson, Sally (2008*) The Higgs Hunter's Guide.* Westview Press.

Hadhazy, Adam (2016, December*) Nothing Really Matters.* Discover Magazine.

Hameroff, Stuart (2008). *"That's life! The geometry of π electron resonance clouds".* In Abbott, D; Davies, P; Pati, A. Quantum aspects of life (PDF). World Scientific. pp. 403–434.

Hatch, Ronald (1995) *"Relativity and GPS,"* Part I, Galilean Electrodynamics, 6, 3, pp. 51-57, and Part II, Ibid. 6, 4, pp. 73-78)

Hatch, Ronald R (1995) *Relativity and GPS*, Part II, Galilean Electrodynamics 6, 4 , p. 73-78/ http://aetherforce.com/the-suppression-of-inconvenient-facts-in-physics-by-rochus-boerner/

Hatch, Ronald (2004) *"Those Scandalous Clocks."* GPS Solutions: 67-73, p. 72.

Hayden, Howard and Whitney, Cynthia (1990) *"If Sagnac and Michelson-Gale, Why Not Michelson-Morley?"* Galilean Electrodynamics, Vol. 1: Nov/Dec.

Hawking, Stephen (2014) *The Beginning of Time.* Lecture: http://www.hawking.org.uk/the-beginning-of-time.html

Henry, Richard Conn (7 July 2005).*The Mental Universe.* Nature 436, 29 | doi:10.1038/436029a.

Herbert, Nick (1985). Notes from Quantum Reality.
http://www.basicincome.com/bp/quantumreality.htm

Hilgevoord, Jan, ed. (1995) *Physics and Our View of the World.*

Hille, Karl (March 23, 2017) *Gravitational Wave Kicks Monster Black Hole Out of Galactic*

Core. https://www.nasa.gov/feature/goddard/2017/gravitational-wave-kicks-monster-black-hole-out-of-galactic-core

Hill, Paul R. (1995) Unconventional Flying Objects: A Former NASA Scientist Explains How UFOs Really Work (Kindle Locations 1261-1265). Hampton Roads Publishing. Kindle Edition.

Holzner, Steven (2004) *Physics II for Dummies*. Hoboken, NJ: Wiley Publishing, Inc.

Holmbberg, Eric (2006) *The Anthropic Principle*. Video produced by BBC Worldwide Ltd.

Hossenfelder, Sabine (2017) *No, physicists have not created "negative mass*. Backreactions blog. http://backreaction.blogspot.com/2017/04/no-physicists-have-not-created-negative.html

Hughes, R.I.G. *The Structure and Interpretation of Quantum Mechanics*. Harvard University Press.

IANDS (International Association for Near Death Studies) (2016) Report of AWARE study. https://iands.org/news/news/front-page-news/1060-aware-study-initial-results-are-published.html

Iowa State Dept. of Physics and Astronomy (2001) *Polaris Project*. http://www.polaris.iastate.edu/EveningStar/Unit2/unit2_sub1.htm.

IPN Progress Report 42-159 2004

Iverson K.E. (1980) "*Notation as A Tool of Thought*", Communications of the ACM, 23: 444–465.

Ives, H. E.; Stilwell, G. R. (1938). "An experimental study of the rate of a moving atomic clock". Journal of the Optical Society of America. 28 (7): 215.

Jaffe, R. (2005). "Casimir effect and the quantum vacuum". Physical Review D. 72 (2): 021301.

Jagerman, Louis (2001) *The Mathematics of Relativity for the Rest of Us*. Victoria, B.C.: Trafford Publishing.

Jansen, K.L.R. (1999) Ketamine (K) and Quantum Psychiatry. Asylum 11 (3) 19-21.

Jorlunde Film Denmark (1985) *Quantum Entanglement Documentary - Atomic Physics and Reality*. Published on youtube in 2014 by Muon Ray.
https://www.youtube.com/watch?v=BFvJOZ51tmc&list=PLtnb8DfCuFNx8x7Jga7K7Wni-eccc3r42

Kaiser, David (2014, Nov. 14). *Is Quantum Entanglement Real?* New York Times Sunday Review: SR10.

Kaku, Michio (2002) Statement on Space as Nothing. Discovery Science Video and (2011) *How the Universe Works*. Video by Discovery Communications produced by Pioneer Productions.

Kaku, Michio (2011) *Michio Kaku Explains String Theory*. Youtube video: https://www.youtube.com/watch?v=kYAdwS5MFjQ

Kaku, Michio (2013) *Is God a Mathematician?* Youtube Video: Big Think Channel https://www.youtube.com/watch?v=jremlZvNDuk.

Kaku, Michio (2015) *Can universes form from "nothing"?* Youtube video: https://www.youtube.com/watch?v=JlcHMI0cC00

Kelly, Alphonsus G. PhD (2005) *Challenging Modern Physics*. Boca Raton, FL: Brown Walker Press.

Kenyon, Dean (2002) Interview in "Unlocking the Mystery of Life." DVD by Illustra Media

Khamehchi, M. A. et al (2017). *Negative-Mass Hydrodynamics in a Spin-Orbit–coupled Bose-Einstein Condensate*, Physical Review Letters DOI: 10.1103/PhysRevLett.118.155301

Kim, Y.H. & Shih, Y. (1999). "Experimental realization of Popper's experiment: violation of the uncertainty principle?" *Foundations of Physics* 29 (12): 1849–1861.

Known Universe (2009) *The Fastest*. Season One: Episode 3. National Geographic Production.

Kolata, Gina (1987) *The Sad Legacy of the Dalkon Shield*. The New York Times: December 6, 1987.

Krasnitz, Michael (2002). *Correlation functions in supersymmetric gauge theories from supergravity fluctuations hHKtions* (PDF). Princeton University Department of Physics: p. 91.

Krauss, Lawrence M. (2012). *A Universe from Nothing: Why There Is Something Rather Than Nothing*. New York: Free Press.

Kuhn, Thomas S. (2012) *The Structure of Scientific Revolutions*. Chicago, IL: University of Chicago Press.

Laureyssens, Dirk (2009). *The Gravitational ETHER of Einstein*. Retrieved from: http://www.mu6.com/einstein.html).

Lee, Penny (1996), "*The Logic and Development of the Linguistic Relativity Principle*", the Whorf Theory Complex: A Critical Reconstruction. John Benjamin's Publishing, p. 84.

Lincoln, Donald (March 13, 2018) *Twin Paradox: the real explanation.* Fermi Lab Youtube site: https://www.youtube.com/watch?v=GgvajuvSpF4

Lincoln, Donald (May 21, 2013) *What is Supersymmetry?* Fermi Lab Youtube site: https://www.youtube.com/watch?v=0CeLRrBAI60

Leplin, Jarrett (1984), *Scientific Realism*, University of California Press.

Lévi-Strauss, C. (1967). *Structural Anthropology*. Translated by Claire Jacobson and Brooke Grundfest Schoepf. New York: Doubleday Anchor Books.

Long, Jeffrey (2010) *Evidence of the Afterlife: The Science of Near-Death Experiences.* New York: HarperCollins.

Maglab (2014) https://nationalmaglab.org/education/magnet-academy/watch-play/interactive/electromagnetic-deflection-in-a-cathode-ray-tube-i

Mandelbaum, Ryan F. (4/21/2017) *No, Scientists Didn't Just Create Negative Mass or Defy the Laws of Physics.* http://gizmodo.com/no-scientists-didnt-just-create-negative-mass-or-defy-1794525465

Markey, Sean (October 8, 2003) *Universe is Finite, "Soccer Ball"-Shaped.* National Geographic News: http://news.nationalgeographic.com/news/2003/10/1008_031008_finiteuniverse.html

Martineau, Harriet (1853) *From the Positive Philosophy of Auguste Comte* (translated London) Vol. I.

Masten, Luke (2006) "Fred Hoyle." In: *The Physics of the Universe.* http://www.physicsoftheuniverse.com/scientists_hoyle.html

Maybury-Lewis, David (1992) *Millennium: Tribal Wisdom and the Modern World*. The Global Television Network: video and book.

Mbarek, Saoussen, Paranjape, M. B. (2014) *Negative mass bubbles in de Sitter space-time*. Phys. Rev. D 90, 101502(R).

Mike, John (2011) *The Anatomy of a Flying Saucer* (Kindle Location 1907). . Kindle Edition.

Milonni, Peter W. (1994) *The Quantum Vacuum: An Introduction to Quantum Electrodynamics* Academic Press.

Minick, Scot (2002) "Unlocking the Mystery of Life". Video by Illustra Media.

Minutephysics (Sep 13, 2017) *Bell's Theorem: The Quantum Venn Diagram Paradox.* https://www.youtube.com/watch?v=zcqZHYo7ONs&list=PLZ7qfdXfdJe6_0ezTgUZD2d9MBRnFAW8D

Mitchell, William C. (2002) *Bye, Bye Big Bang, Hello Reality.* Carson .City, Nevada: Cosmic Sense Books.

Moon, P. and Spencer, D.E. (1956) "*On the Establishment of Universal Time*", Phil. Sci., Vol. 23, No. 3 (Jul., 1956), pp. 216-229.

Musser, George (2003) *According to the big bang theory, all the matter in the universe erupted from a singularity. Why didn't all this matter--cheek by jowl as it was--immediately collapse into a black hole?* Scientific American: scientificamerican.com/article/according-to-the-big-bang/

Munday, J.N.; Capasso, F.; Parsegian, V.A. (2009). "Measured long-range repulsive Casimir-Lifshitz forces". Nature. 457 (7226): 170–3.

Naeser, C. W. (1979). *Fission-Track Dating and Geologic Annealing of Fission Tracks.* In: Jäger, E. and J. C. Hunziker, Lectures in Isotope Geology, Springer-Verlag.

Nave C. R. (2016) "*Electroweak Unification*". Hyperphysics (Georgia State University). http://hyperphysics.phy-astr.gsu.edu/hbase/Forces/unify.html

Navipedia: Troposphere Monitoring. www.navipedia.net

"NAVSTAR GPS User Equipment Introduction" (PDF). US Coast guard navigation center. US Coast Guard. September 1996

Nawrot, W. (1998) "*Some remarks on Around-the-World Atomic Clocks Experiment*", Submitted to International Journal of Theoretical Physics.

Ornstein, Robert (1977) *The Psychology of Consciousness.* New York: Harcourt, Brace, Javanovich.

Ornstein, Robert (1997) *The Right Mind.* New York: Harcourt, Brace, & Company.

Orzel, Chad (2018) *The Real Reasons Quantum Entanglement Doesn't Allow Faster-Than-Light Communication.* https://www.forbes.com/sites/chadorzel/2016/05/04/the-real-reasons-quantum-entanglement-doesnt-allow-faster-than-light-communication/#699818543a1e

Pandian, Jagadheep D. (June 27, 2015) "Why is the Universe flat and not spherical?" (Advanced)

Ask an Astronomer: http://curious.astro.cornell.edu/about-us/103-the-universe/cosmology-and-the-big-bang/geometry-of-space-time/600-why-is-the-universe-flat-and-not-spherical-advanced

Parapsychological Association (February 11, 2011) *PK on random number generator.* http://www.parapsych.org/articles/36/66/1_pk_on_random_number_generators.aspx

Paul (2016) *Michelson-Morley Experiment Explained.* Youtube video: https://www.youtube.com/watch?v=F2m_VZJM0Zc

Penrose, Roger (1995). *Shadows of the Mind: A Search for the Missing Science of Consciousness.* Oxford: Oxford Univ. Press.

Penrose, Roger (1999). *The Emperor's New Mind: Concerning Computers, Minds, and the Laws of Physics.* Oxford: Oxford Univ. Press.

Petkov, Vesselin (14 December 2003) *Propagation of light in non-inertial reference frames.* Science College, Concordia University. Retrieved from: https://arxiv.org/pdf/gr-qc/9909081.pdf

Pew, Glen (2010). Supersonic Flight, Sonic Booms. AVweb Video: Retrieved from Youtube: https://www.youtube.com/watch?v=gWGLAAYdbbc.

Physics Forums (Oct.19, 2012) https://www.physicsforums.com/threads/do-gravity-waves-imply-repulsive-force-component.645331/

Pike, O, J. et al. (2014) 'A photon–photon collider in a vacuum hohlraum'. *Nature Photonics*, 18 May 2014.

Pogge, Richard (2015) *Real-World Relativity: The GPS Navigation System.* http://www.astronomy.ohio-state.edu/~pogge/Ast162/Unit5/gps.html

Popper, Karl (1962). *Conjectures and Refutations.* New York: Harper Torchbooks.

Popper, K. R. (1967), "Quantum Mechanics Without 'the Observer'", in Mario Bunge (ed.), Quantum Theory and Reality, New York: Springer, pp. 1–12.

Popper, Karl (1982).*Quantum Theory and the Schism in Physics.*

Popper, Karl (1985). "Realism in quantum mechanics and a new version of the EPR experiment" In Tarozzi, G.; van der Merwe, A. Open Questions in Quantum Physics. Dordrecht: Reidel. pp. 3–25.

Powell, Kevin (2016, July/August) *Entanglement.* Discover

Quigg, Chris (2008, January 17). "Sidebar: Solving the Higgs Puzzle". Scientific American.

Reucroft, Stephen and Swain, John (2016) *What is the CASIMIR EFFECT?* Scientific American website, retrieved 12-30-2016, https://www.scientificamerican.com/article/what-is-the-casimir-effec/#checkout

Ricker, Harry H. III *What Happened to Dingle?* http://www.mrelativity.net/Papers/18/Ricker.htm

Sears, Young & Zemansky (1999) *University Physics*. Addison-Wesley. Pp.18–38.

Shankland, R.S. et al. (1955) Rev. Mod. Phys. 27 no. 2, pp. 167–178

Sheldrake, Rupert (2009) *Morphic Resonance*. Rochester: Rock Street Press.

Shifman, Mikhail (October 31, 2012) *Reflections and Impressionistic Portrait at the Conference Frontiers: Beyond the Standard Model*, FTPI (pdf).

Shockey, Peter (2013) *George Rodonaia's – NDE – A Scientist's Afterlife.* Documentary: http://www.lifeafterlife.tv/

Smolin, Lee (2006) *The Trouble with Physics: The Rise of String Theory, The Fall of Science, And What Comes Next.* Harcourt Brace Publishers

Sofaer, A., Zinser, V. & Sinclair, R. M. (1979 a). *A Unique Solar Marking Construct.* Science, 206, 283-291.

Sokal, Alan D. (5 June 1996). "A Physicist Experiments with Cultural Studies". *Lingua Franca*.

Sorensen, Eric (April 17, 2017) *Physicists create 'negative mass.'* https://phys.org/news/2017-04-physicists-negative-mass.htm

Sowell, Thomas (1995) "*Ethnicity and IQ*". In Steven Fraser, ed., *The Bell Curve Wars.* New York: Basic Books, pp. 70-79.

SI Brochure. BIPM (December 22, 2013) Unit of time (second).

Spencer, Domina Eberle & Shama, Uma. *A New Interpretation of the Hafele-Keating Experiment.* http://www.shaping.ru/congress/english/spenser1/spencer1.asp

Stacks Physics Exchange: (2012) http://physics.stackexchange.com/questions/44934/does-matter-with-negative-mass-exist

Styer, Daniel F. (2011) *Relativity for the Questioning Mind.* Baltimore: Johns Hopkins University Press.

Susskind, Leonard (7 July 2008). *The Black Hole War: My Battle with Stephen Hawking to Make the World Safe for Quantum Mechanics*. Hachette Inc.

Susskind, Leonard (Jul 16, 2013) *Why is Time a One-Way Street?* Lecture at Santa Fe Institute. https://www.youtube.com/watch?v=jhnKBKZvb_U

Taylor, John (1980). *Science and the Supernatural: An Investigation of Paranormal Phenomena Including Psychic Healing, Clairvoyance, Telepathy, and Precognition by a Distinguished Physicist and Mathematician*. London: T. Smith.

Talbot, Michael (1992) *The Holographic Universe*. New York: Harper Collins.

Templeton, Graham (November 19, 2012) *Stanford's quantum entanglement device brings us one step closer to quantum cryptography*. Extreme Tech: www.extremetech.com/extreme/140739-stanfords-quantum-entanglement-device-brings-us-one-step-closer-to-quantum-cryptography

Tesla, Nikola (1932) *Space is Nothing*. New York Herald Tribune (September 11)

The Physics Classroom. *The Forbidden F-Word* (1996-2016). http://www.physicsclassroom.com/class/circles/Lesson-1/The-Forbidden-F-Word

Thornbill, Wallace and Talbott, David (2007) *The Electric Universe.* Portland, OR: Mikamar Publishing.

Trubody, Ben (June, 2017) "Richard Feyman's Philosophy of Science." *Philosophy Now*, Issue 120. https://philosophynow.org/issues/114/Richard_Feynmans_Philosophy_of_Science

Tylor, Edward B. (1877) *Primitive Culture.* New York: Henry Holt.

Tytell, David (April 13, 2004) "Building Planets In Plastic-Bags" http://www.skyandtelescope.com/astronomy-news/building-planets-in-plastic-bags/

University of Illinois Physics Website (2007) *Why are atomic masses not expressed as whole numbers?* https://van.physics.illinois.edu/qa/listing.php?id=1216

Veritasium YouTube site (2016) *The Absurdity of Detecting Gravitational Waves*. Interview with Prof Rana Adhikari. https://www.youtube.com/watch?v=iphcyNWFD10&feature=youtu.be

Vessot, R. F. C. and Levine, M. W. (1979, Dec.). Gravitational Redshift Space-Probe Experiment GP-A Project Final Report Contract NAS8-27969 Retrieved from http://ntrs.nasa.gov/archive/nasa/casi.ntrs.nasa.gov/19800011717.pdf.

Von Baeyer, Hans Christian (March 8,2016) "Quantum Weirdness? It's All in Your Mind" in *Physics at the Limits*, Scientific American Publication.
Readers Respond to "Quantum Weirdness": https://www.scientificamerican.com/article/readers-respond-to-quantum-weirdness/

Wall, Mike (March 24, 2017) *Gravitational Waves Send Supermassive Black Hole Flying.* SPACE.com

Wasson, Valentina and Wasson, Gordon (1957) *Mushrooms, Russia and History.* New York: Pantheon, Vol. 2, 264-65.

Weber, Renee (1982) *The Holographic Paradigm* (Fritjof Capra interview) Shambhala/Random House, pp. 217–218.

Wolchover, Natalie (06.30.14) Have We Been Interpreting Quantum Mechanics Wrong This Whole Time? *Quanta Magazines.* https://www.wired.com/2014/06/the-new-quantum-reality/

Wigner, E.P. (1961), "Remarks on the mind-body question", in: I.J. Good, *The Scientist Speculates*, London, Heinemann.

Wikipedia (retrieved 2017) Faster-*than*-Light. https://en.wikipedia.org/wiki/Faster-than-light.

Wikiquotes: https://en.wikiquote.org/wiki/William_Thomson

Woit, Peter (2006). *Not Even Wrong: The Failure of String Theory and the Search for Unity in Physical Law.* Basic Books.

Yaglom, Isaak Moiseevich (1980) Mathematical Structures and Mathematical Modelling. Philadelphia: Gordon and Breach Science Publishers.

White, Leslie (1956) The Locus of Mathematical Reality. In: The World of Mathematics by Newman, James R. New York: Murray Printing Company.

Zato Tomáš (Jun 17, 2014) *What-Is-Hawking-Radiation-And-How-Does-It-Cause-A-Black-Hole-To-Evaporate.* Stack Exchange.
https://physics.stackexchange.com/questions/26605/what-is-hawking-radiation-and-how-does-it-cause-a-black-hole-to-evaporate

Zukav, Gary. (1979) *The Dancing Wu Li Masters: An Overview of the New Physics.* HarperCollins. Kindle Edition.